BROWSE

NUTRITION

BROWSE
NUTRITION

SEMIARID REGIONS

ROQUE RAMIREZ LOZANO

To order additional copies of this book, please contact:
Palibrio
1663 Liberty Drive
Suite 200
Bloomington, IN 47403
Toll Free from the U.S.A 877.407.5847
Toll Free from Mexico 01.800.288.2243
Toll Free from Spain 900.866.949
From other International locations +1.812.671.9757
Fax: 01.812.355.1576
orders@palibrio.com
732032

CONTENTS

PRESENTATION

Browse is the tender shoots, twigs or leaves of trees or shrubs that are acceptable for grazing. Browse plants, beside grasses, constitute one of the cheapest sources of feed for ruminants. Browse plants provide vitamins and very frequently mineral elements, which are mostly lacking in grassland pastures. Moreover, browse species are indispensable sources of animal feed in the world, particularly in areas with dry to semidry climates. Such species can alleviate feed shortages or even fill feed gaps in the winter and especially in the spring, when grassland growth is limited or dormant due to unfavorable weather conditions. These include several spontaneous shrubs and trees, which are essential components of natural communities such as shrublands and woodlands. They cover large areas and constitute grazing lands for all domestic animals, mainly goats, deer and sheep. Browse production and nutritive value vary widely among species and varieties.

This book, in 45 chapters, describes the nutritional potential of noticeable trees and shrubs growing in northeastern Mexico and southern Texas, USA. It analyses the browse utilization by range ruminants. The author argues the importance browse minerals as food resources. He debates the plant mechanisms against herbivory. He describes the leaf litter as a feed resource for range small ruminants. The author examines the diet selection by ruminants. In addition, this book covers the taxonomic characteristics and nutritional profiles of the foliage of 18 legumes and 21 nonlegumes native tree/shrub species growing in the semiarid regions of Northern Mexico.

The Author

Prof. Roque Gonzalo Ramírez Lozano, Ph.D.
Universidad Autónoma de Nuevo León
Facultad de Ciencias Biológicas, Alimentos,
Ave. Pedro de Alba y Manuel Barragán S/N,
Ciudad Universitaria, San Nicolás de los Garza,
Nuevo León, 66455, México.
e-mail: roque.ramirezlz@uanl.edu.mx

This book is dedicated in memory to my mother: Ernestina Lozano de Ramírez (1926-1993) for her endless support ......

CHAPTER 1

Foliage from trees and shrubs as a feed resource

Introduction

Browse is the tender shoots, twigs or leaves of trees or shrubs that are acceptable for grazing. Browse plants, beside grasses, constitute one of the cheapest sources of feed for ruminants. Browse plants provide vitamins and very frequently mineral elements, which are mostly lacking in grassland pastures. Moreover, browse species are indispensable sources of animal feed in the world, particularly in areas with dry to semidry climates. Such species can alleviate feed shortages or even fill feed gaps in the winter and especially in the spring, when grassland growth is limited or dormant due to unfavorable weather conditions. These include several spontaneous shrubs and trees, which are essential components of natural communities such as shrublands and woodlands. They cover large areas and constitute grazing lands for all domestic animals, mainly goats, deer and sheep. Browse production and nutritive value vary widely among species and varieties. The nutrients vary greatly according to season, with a higher concentration of fiber and ash. Their nutritive value, however, does not always relate to their chemical composition due to the presence, in most species, of secondary compounds of plants such as tannins, alkaloids, saponins, essential oils and oxalates, which limit nutrient utilization and reduce animal performance.

Browse utilization by ruminant animals

Acknowledgement of the potential use of tree and shrub foliage to produce substantial quantities of high protein biomass has directed to the expansion of animal agriculture systems, which incorporate the use of these forages with local coarse food resources. With the purpose to estimate the suitability of trees and shrubs as components of ruminant fiber rations, information is essential in many aspects, including:

1. Capability and capacity of the trees and shrubs to regenerate foliage when grazed or harvested
2. Feeding behavior of animals when challenged with tree and shrub foliage
3. Voluntary intake of trees and shrubs foliage under diverse environmental situations
4. Adaptation of trees and shrubs to the native situations and their potential to convert in weeds
5. Ease of seedling formation, level of growth and renewal
6. Growth tendency of trees/shrubs in relation to crops or grass
7. Required soil pH features and nutrient grade
8. Nutritional value of the forage and its alteration with mowing, browsing or farming.

It also has been recognized that besides from their role as animal feeds, trees and shrubs are valuable sources of:

1. Chemical compounds with pharmacological properties
2. Cycle nutrients through leaf fall
3. Defend soil from water and wind erosion
4. Fuel wood
5. Green manure or mulch
6. Food for humans (mainly fruits)
7. Landscape improvement
8. Offer employment
9. Produce income
10. Shelter and shadow
11. Store carbon, reducing the Greenhouse Effect

12. Timber and fencing materials
13. Wildlife habitat

Of the above, the most important issues for small farmers in developing countries are food, feed, fuel wood and timber. The order of importance of these depends on the particular country or region. It is now generally recognized that browse fodder is a potential inexpensive locally produced protein supplement for ruminants, and can correct nitrogen deficiency in herbaceous vegetation, especially during the dry season.

In the past, shrubs (browse) species were considered poor feeds for animals and efforts were made to control or eradicate them from grasslands. Over the last 25–30 years this attitude has changed and substantial research has been completed to show that woody species are important forage resources in the world. However, a number of nutritional problems have been identified in several species. An interesting challenge for scientists in the field of animal nutrition is the introduction of alternative feedstuffs such as edible shrubs that could overcome the problems of environmental harshness and production costs. However, the feeding value of shrubs is widely variable, depending on species or cultivars, plant parts, phenological stage, environmental conditions and management. Moreover, its digestibility is low due to the high presence of secondary compounds, especially lignin and tannins. Nevertheless, deciduous species maintain their nutritional value during harsh seasons and can be used as feed supplements for ruminants. In addition, the use of browse plants seems to be vital for livestock farmers, especially during the hardest period of the year. Solutions have to be found to improve forage availability all the year around for the poor agropastoralists, while ensuring environmental sustainability.

Ruminant animals as browsers, intermediate feeders, or grazers. Moose and deer are known as browsers. These animals consume a diet largely consisting of highly digestible forbs (broad-leaved weeds and legumes) and browse (leaves from woody plants). Grazers like bison and cows comprise of ruminants, which eat a great amounts of grasses for their nutrition. Goats and elk are intermediate feeders and are opportunistic feeders that will change the diet selectivity between browse and grasses depending to food availability and palatability. The ruminant animals have physical variations in some body aspects counting rumen construction, tongue, mouth and teeth to permit them to more efficiently develop their selected diet. For instance, deer has a minor pointed muzzle with a thin tongue to support in gaining a greatly selected diet of browse and forbs. Small ruminants, in addition, have a rumen constitution to enable a quickly assimilated diet and a bigger liver to process tannins that sometimes happen in browse plants. Cows, however, have a wide plane muzzle, flat teeth for crushing diets high in grass fiber, and a very great rumen to house the extra time required to course a slowly fermented diet.

Browse in ruminant productivity

Integration of shrub and tree species in livestock production systems may be a feasible option to increase the use of rangelands and to expand the dietary options by ruminants. Several shrubs and trees are simply proliferated and they not necessitate a great level of management practices. Furthermore, some browse plants have levels of crude protein (CP) that are bigger than other foods usually utilized in animal production that increases consumption of fibrous feeds by ruminant animals. Browse plants can have good entree to water that infiltrates through the upper soil, perchlorates into the subsoil (30–150 cm), and may have the ability to yield high amount and quality feed in places with extended drought times. Several browse species have a prolonged life.

The difficulty of ruminant food supply and quality is intensified in semi-arid tropical regions and arid with unpredictable and variable rainfall that restricts biomass production and developing period of herbaceous plants. Therefore, ruminant animals in these areas have to live on reduced food resources of low nutritional quality for most of the time. For instance, CP content of browse plants in the rainy periods was described to be rational, but fall radically during the drought period in semiarid regions, leading to extended period of low nutrition of animals remaining in such situations. Moreover, unrestrained utilization of progressively uncommon communal browsing regions on drought rangelands has augmented to their poverty; furthermore, diminishing the accessibility of animal food resources. The additional of difficulties associated to animal feeding in drought regions need manipulating potential animal food sources like browse plants that are adjusted to these conditions. Native browse plants are valuable resources of animal foods, mainly during the drought period because of the residual green and providing browse plants with good nutritive quality when all other annual plants such as grasses and forb species have expired. In addition, browse species such as the legumes that trap nitrogen from the atmosphere, and this provides to their augmented CP concentration.

Results from a study on *Acacia* species showed that the CP of foliage is high enough to use as a supplement to low quality diets, and that species are high in all mineral elements. Acacia species incline to keep green extended though the dry period and aid as resource of nutrients required for small ruminants. Nonetheless, Acacia plants have the difficulty of having phenolics that comprise tannins. These compounds have an adverse consequence on the food quality of the plant, thus upset consumption and digestion of the ingesta. Tannins may bind to protein compounds and later yield them inaccessible for use by the ruminant. Thus, when leaves that are rich in tannins and are offered as an only feed to ruminants they cannot offer the maintenance needs regardless of their comparatively high CP concentration and low fibrous concentration. Tannins can also have adverse properties on milk yield, wool growth and weight gain of ruminants.

The utmost value of foliages from browse plants and particularly trees is their function as dietary complements. This function is related to their source of diet protein, vitamins, minerals and energy. In many parts of the world, there are systematic food scarcities and lack of rainfall. In those situations, survival feeding, mostly on cereal stubbles, consequences in diminished live weight gain and continuous low performance of the livestock. The use of foliage supplements is another approach that maybe has not given satisfactory investigation and development consideration. For some aspects, this consideration, more than any other, has huge prospective for claim with ruminants, especially in conditions where the livestock are copious and diverse. It is suitable, thus, to evaluate the present empathetic of the utilization of foliage complements and the profits of this approach.

Chemical composition

The value of browse plants as CP rich foliage for improving livestock performance has been undervalued. In northeastern Mexico, the fodder from browse plants denotes an important quantity of accessible feed for browsing small ruminants, and in some aspects is the unique resource of food nutrients. In developing regions of the world, establishing browse plants and trees and shrubs for animal feeding is a common management system that has been rising recently. Many browse species that produce in this region have been utilized as foliage for small ruminants, particularly leguminous species. In other semiarid areas of the world, the use of trees and shrubs has been encouraged because of severe feed scarcity, harsh and continued dry times, low livestock production, death and inefficient production. Nevertheless, the low value and seasonal environment of the foliage source, along with low consumption by livestock and low digestion of foliage, are the main aspects causal to the low performance of animals provided by browse plants.

Native plants in the semiarid areas of northeastern Mexico are significant food sources for free range small ruminants. The nutritional profile was been evaluated alone with the rate and extent of nutrient digestion of browse plants that grow in northeastern México and that are ingested by free range small ruminants. The *Medicago sativa* hay was used as reference forage of high nutritional quality. The organic matter content fluctuated in a small but significant range during the year. Moreover, in general, all plants had more CP during spring and summer seasons, and were lower in autumn and winter seasons. Browse plants had higher CP at the end of the spring and at the start of the autumn. It has been recognized that in numerous areas of the world, CP concentration in browse species is elevated, and it is comparatively continuous during the year, compared to other plant species such as grasses. It has also been recognized that browse species with low fiber concentration have an elevated nutritional value and young foliages have lesser fiber.

In addition, it has been informed that cell wall (CW) in all evaluated browse species was lower to that of *Medicago sativa* hay. This circumstance may signify a benefit for animals that are alimented with these forages, due to foliages, which have low CW may be fermented without difficulty and in more measurement. The maximum amount of lignin in the plants happened in spring and summer seasons. Most browse species had an annual average value of lignin higher that *Medicago sativa* hay. Of all evaluated plants, four had trace or very small quantities of tannins. It seems that the levels of tannins, met in the assessed plants in this study, were not toxic for small ruminants.

In other study, leaves from leguminous plants such as *Havardia pallens, Parkinsonia aculeata, Eysenhardtia polystachya, Pithecellobium ebano, Caesalpinia mexicana,* and non-leguminous such as *Celtis pallida, Bernardia myricaefolia, Helietta parvifolia Gymnosperma glutinosum,* and *Diospyros texana* were nutritionally evaluated. During the summer, leaves of all plants had higher CP concentrations, however, during autumn and winter seasons were lower. *Medicago sativa* hay was involved as comparative food with high CP concentration (22.0%). Species such as *Pithecellobium ebano, Eysenhartia polystachya, Celtis pallida, Havardia pallens* and *Helietta parvifolia,* had CP values similar and in some of them with substantial CP than *Medicago sativa* hay. The CP from browse plants is often reviewed as a CP complement for both wildlife and livestock.

There is an extensive margin in CP value among trees and shrubs. The average figure of 277 browse plants revised from 22 literature scientific papers exhibited a figure of 17.0%, and they varied from 11.1 to 41.7%. Most of assessed browse species had low CW concentration throughout the year. Exception was for *Parkinsonia texana* and *Havardia pallens* that had high in CW value. These species had higher CW value than *M. sativa* hay. Low CW and thus elevated cell content make these species with elevated nutritional quality compared to grass species. In this investigation, during the winter most browse plants had low CW values, conversely, during summer the CW content was higher. Low cellulose content was reported in all species in winter, but in summer, browse plants were higher. *Parkinsonia aculeata* was highest in cellulose content. During all seasons, hemicellulose value was lower than cellulose. It was argued that foliage from temperate leguminous plants were lower hemicellulose than cellulose; nonetheless, hemicellulose and cellulose were similar in warm plants.

The chemical composition was evaluated of seven shrubs that grow in northeastern Mexico. They reported that CP content was significantly different among seasons and were higher in winter. High CP ranges was found in plants such as *Parkinsonia texana, Acacia berlandieri, Leucaena leucocephala, Desmanthus virgathus* and *Acacia greggii.* Plants such as *Acacia amentacea* and *Ziziphus obtusifolia* had CP values comparable to M. sativa hay. They also reported that most of the evaluated plants had low cell wall content mainly during winter and spring seasons, but in summer and fall were high. With exception of *Acacia amentacea*, all shrubs had annual mean CW content lower than *M. sativa* hay. Leucaena leucocephala had cell wall values comparable to M. sativa hay. Cellulose values were low during summer and fall, but in spring and winter, plants were high in cellulose. The hemicellulose content in leaves of all plants was low during summer and fall, but it was higher in spring and winter. During winter and spring lignin content was relatively low in all evaluated plants, but in summer and fall lignin content was rather high. Condensed tannins (CT) in evaluated plants were significantly different among seasons. Exception was *Ziziphus obtusifolia* that had similar tannin content in all four seasons. *Acacia berlandieri* had the highest seasonal ranges and *Parkinsonia aculeata* and A*cacia greggii* had the lowest. During winter, all plants had low condensed tannins, but in summer were high.

It has been reported that five legumes species (*Acacia peninsularis, Parkinsonia texana, Pithecellobium confine, Prosopis* sp. and *Mimosa xantii*) and five non legumes species (*Bursera microphyla, Cyrtocarpa edulis, Lippia palmeri, Opuntia cholla* and *Turnera diffusa*) growing in a northwestern region of Mexico, the legumes constituted a better quality forage source because to almost double CP concentration and degradable CP content than nonlegumes, especially during the worst quality forage season (spring and summer). In addition, legumes maintained a good quality during the year and the year variation may be more important in determining forage quality. Species such as *Cercidium floridum* and *Prosopis* sp. had the best quality protein. The nonlegumes species have a seasonal effect on CP composition and degradation; these variations make them a poor forage quality source during spring and summer, and only after the rainy season, in autumn, nonlegumes species may be of better quality, especially *Cyrtocarpa edulis*. The authors also stated that legumes were highest in nutritive quality during the dry season (spring and summer). They provided a higher available CP and fermentable organic matter (OM) through the year. Non-legumes were more variable across seasons in degradation characteristics. Within legumes species *Pithecellobium confine* was lowest in nutritive quality due its high CW and lignin content. *Cercidium floridum* and *Prosopis* sp. were highest because they provided more available OM and CP, even when *Cercidium floridum* was high in tannins. *Opuntia cholla* was a good fermentable OM source because it was high in hemicellulose. The authors concluded that all forage species had a lower Zn and Cu content than beef cattle requirements, due their seasonal variation in Mn and P content; they may be lower in these minerals during the dry season or years of drought. *Bursera microphylla* and *Cyrtocarpa edulis* were a good source of Na, and *Bursera microphylla*, *Opuntia cholla* and *Turnera diffusa* of Mn. Lignin, tannins, NDF and cellulose content limited OM and CP degradation in legumes and non legumes species. Some minerals under beef cattle requirements concentrations were met, they also may limit OM and CP degradation (Zn, P), others minerals may increase it (K and Ca).

In a region of Durango state, Mexico, the nutritional value was reported in the following groups of native plants: trees (2), shrubs (12), forbs (4), cacti (3), flowers, fruits and pods (8) that are consumed by range goats. It was reported that species varied in their contents (% dry matter) of ash (4.0-32), CP (3.9-18.6), CW (18.7-65.1), lignin (1.1-14.5) and CT (0.1-9.1). Levels of Ca, Mg and Fe in species may be adequate to meet goats requirements, while P, K and Mn could be adequate in 90% of the plants; 59% of the plants were deficient in Na, Cu and Zn. A wide variability in in vitro gas production parameters, ME and VFA concentration was also recorded. Only 28% of the studied species would satisfy ME of goat requirements. A positive effect of polyethylene glycol (PEG) on in vitro gas production was evident in tree and shrub species. In vitro fermentation parameters (b and c) and the concentration of volatile fatty acids were negatively correlated with CW and lignin, which might explain the detrimental effect of CW compounds on in vitro gas production. Values from the *in vitro* gas production parameters, mineral content, in vitro dry matter (DM) digestibility, in vitro organic matter (OM) digestibility, in situ CP degradability and metabolizable protein, support the nutritional relevance of shrubs, cacti, forbs and pods for grazing goats in the semiarid regions of North Mexico. Nonetheless, supplementation program might be required to provide with degradable protein to the goats when consuming shrubs, trees, cacti and fruits and pods. Similarly, fermentable energy would be advised in periods when goats are to consume forbs. Goats consuming cacti might require as well P supplementation; whereas, Mn ought to be offered when the animals select fruits and pods. Meanwhile, Na, Cu and Zn supplementation is necessary all year around.

In a sarcocaulescente thornscrub vegetation at La Paz county, of the state of Baja California Sur, México, it was determined that during eight seasons, beginning in summer 2006 and ending in spring of 2008, the nutrient intake by goats. The author reported that the intake of energy, CP, Ca, K, Fe, Mg and Mn was sufficient to satisfy the requirements of an adult goat of 37 kg of live weight gaining 25 g d-1. However, Cu and Zn were deficient. Variations in chemical composition and digestibility of goat diets and in the nutrient intake by goats could be attributable to the seasonal changes that influenced climatic conditions.

The foliar tissue of browse plants from northeastern Mexico such as *Acacia amentacea, Bumelia celastrina, Croton cortesianus, Karwinskia humboldtiana, Leucophyllum frutenscens* and *Prosopis laevigata* was evaluated seasonally (from August 2004 to May 2006), during two consecutive years, the Ca, K, Mg, Na, P, Cu, Fe, Mn and Zn contents in. It was reported yearly and seasonal variations in plant minerals, and in spite of these differences, all plant species had suitable levels of Ca, Mg, K, Fe and Mn to satisfy grazing ruminant requirements. However, P, Na, Zn and Cu, showed marginal inadequate concentrations in prolonged periods throughout the year and it might have a negative impact on animal productivity. Similar responses were found with other browse plants (*Castela texana, Celtis pallida, Forestiera angustifolia, Lantana macropoda, Zanthoxylum fagara*) collected in the same region and at same time. It was registered that regardless of spatio-temporal differences, all plant species had suitable levels of Ca, Mg, K, Cu, Fe and Mn to satisfy range domestic and wild ruminant requirements. Nonetheless, P, Na and Zn showed marginal inadequacies in some seasons throughout the year.

It was reported the monthly chemical composition of the diet consumed by range goats and plasma concentrations of glucose, urea, non-esterified fatty acids (NEFA), luteinizing hormone (LH), growth hormone (GH), insulin and progesterone (P4) during lactation (January-June), reported that body weight of goats decreased (P<0.05) by 12.5% during the period from January to May, whereas in June goats lost 250 g d-1. Dietary CP (range = 8.0-12.7%), CW (36.8-60.2), lignin (8.2-15.2), hemicellulose (8.4-19.3) and cellulose (7.8-33.3) differed (P<0.05) between months. Plasma glucose concentration (32.0-59.7 mg dl-1), NEFA (0.3-0.9 nM l-1) and urea (12.3-19.9 mg dl-1) varied (P<0.05) among months. With the exception of progesterone (0.1-0.6 ng ml-1), LH (7.1-11.0), GH (22.8-30) and insulin (0.6-1.8) increased (P<0.05) as lactation period progressed. Based on plasma metabolite concentrations, the authors suggested that an energy supplementation schedule might be necessary during the early lactation period of goats. In addition, if harsh climatic conditions appear at the end of lactation, an increment of 70% in their energy maintenance requirements might be considered to avoid weight losses and to improve body condition of goats prior to the breeding season.

Browse plants play an important role in providing feed for white tailed-deer in semiarid rangelands of northeastern Mexico. The Chemical composition and in vitro ruminal fermentation of leaves with or without PEG treatment were determined in browse plants consumed by white tailed deer such as *Acacia amentacea, Celtis pallida, Forestiera angustifolia* and *Parkinsonia texana* in undisturbed rangelands. Leaves were collected at two sampling times (January and April, 2009) in three county sites (China, Linares and Los Ramones) of the state of Nuevo Leon, Mexico. A wide range in chemical composition and in vitro gas production kinetics was observed among sites, species and among sampling times within each species. In plants with higher CT content such as Acacia amentacea (CT = 18%) and Parkinsonia texana (8%), the PEG treatment significantly increased the in vitro gas production parameters and metabolizable energy. *Celtis pallida* had the highest in vitro fermentation parameters and could be due to its lower lignin (ADL) and CT levels. Some variation was observed among shrubs such as in *Forestiera angustifolia* that had lower fermentation and lower ADL and CT. This discrepancy could be due to genotypic characteristics relative to the type of secondary compound activity. The authors concluded that all plants resulted with high nutritional value for white-tailed deer knowing that they have digestive mechanisms to neutralize CT.

In four experiments, the influence of eight shrub leaves on digestion, nitrogen retention and ruminal digestion characteristics by sheep fed low quality based diets. It was found that in all experiments only few variations in nutrient intake, nutrient digestion coefficients and N retention were observed between sheep fed diets containing *M. sativa* hay and sheep fed diets with different levels of shrubs. The same responses were detected in sheep rumen fermentation parameters and data for digestibility values: a, b and c values and effective degradability of dry matter, crude protein and neutral detergent fiber of individual forages. Even though, all shrub species contained high levels of CP, it seemed that plant secondary compounds in browse species affected their nutritional quality, reducing the nutrition of sheep fed diets with different levels of browse plants. In addition, he suggested that evaluated shrubs could serve effectively as protein supplements to poor quality roughages, mainly during the dry season. For its nutritive value, Leucaena leucocephala (even though has mimosine), *Pithecellobium pallens, Celtis pallida* and *Acacia greggii* could be suitable replacement for alfalfa hay and possibly to other commercial supplements, which would lower production costs. *Acacia berlandieri, Parkinsonia aculeata, Ziziphus obtusifolia* and *Prosopis juliflora* not seems to be a replacement option, because its high level of plant secondary compounds such as condensed tannins, saponins, flavonoids and volatile oils among others limited the animal response. However, it is known that the level of these plant detrimental secondary compounds vary according to region, season and part of the plant, making it possible to develop strategies to exploit the high nitrogen content of the shrubs. It was recommended that it is possible that moderate levels of condensed tannins achieving greater efficiency in animal production; moreover, PEG may be added to forage based diets rich in tannins for ruminant feeding because it binds to tannins and thus prevent the formation of potentially indigestible tannin–protein complexes. Moreover, under controlled conditions PEG supplementation dramatically increased the intake and digestibility of some local tannin rich plants

CHAPTER 2

Minerals in foliage from trees and shrubs

Introduction

One of the main problems limiting livestock production in many areas of the world is the nutritional status of the animals. According to most researchers, the main factors limiting the productive behavior of grazing animals are the low protein content of plants, low dry matter intake due to the high fiber content in forages and mineral deficiencies and/or vitamins. However, it must be distinguished that the problems of mineral nutrition refer not only to cases of deficiencies, but also to the toxic levels of some elements such as Hg, Al, Cd, Pb, and even those essential elements such as Cu, F, Mo, and Se that may limit livestock production in certain regions. Moreover, Ca, Cu or Fe in excess, can be detrimental to production compared to any benefit derived from mineral supplementation.

Minerals in soil and plants

With the purpose to provide a mixture of salts and minerals that meet the animal requirements of these elements, it has to know that besides the concentration of minerals in the forage consumed, there are minerals present in water and soil. The mineral nutrition problems are closely linked to the soil characteristics. Poor soils are well-defined geographical areas and the animals that live those grazing may suffer endemic problems. However, it has been noted that the soil analysis is not an exact element of judgment on the mineral nutrition of the animal that lives in these soils. This is because the requirements of the animal are in many cases so small and the soil analysis reveals possible deficiencies in the animal.

Thus, it is more useful to analyze what the animal consumes in each region. Research in this field is limited and has been focused on the study of the mineral concentration in the whole plant, which cannot properly reflect the nutritional value of the plant because cattle generally prefer living plant tissue and dead tissue prefer leaves the stems, so that a more realistic analysis of the mineral content of the diet of livestock can be obtained from living plant tissue.

Factors affecting mineral requirements

The quantity of minerals required will vary depending on the age, weight, health, species and type and level of production of the animal. Young animals absorb minerals such as Ca more efficiently than older animals, but they have higher mineral requirements. High rates of gain or milk production and poor health or parasitic burden will increase mineral requirements. Small ruminants (sheep, goats and white-tailed deer) have different requirements than large ruminants (cattle), with copper being the prime example. For sheep, levels above 25 mg/kg of Cu are considered toxic, while cattle do not reach toxicity levels until 400 mg/kg are present in the diet. As Cu toxicity in sheep is swift and deadly, it is crucial that only minerals designated for sheep be used. Copper supplied in feed usually is adequate and sheep copper supplementation is unnecessary and dangerous. The bottom line is one feed or ration will not meet the requirements of every class of livestock. Rations should be formulated based on chemical analysis of home-grown feeds and rechecked anytime a feed, livestock class or feeding regime is changed; otherwise production may suffer.

Mineral interactions

Feeds that are inadequately combined in a diet will have a marked effect on the quantity and proportions of specific minerals. Although a diet may contain adequate amounts of a certain nutrient, it may be rendered largely unavailable by too much of another nutrient. Ca:P ratios should be kept between 1:1 and 2.5:1 for breeding animals and between 1.5:1 and 3.5:1 for growing finishing beef cattle, goats and sheep to prevent this from occurring. Proportions of trace minerals in diet are also important, for example, high levels of Mo and S have been shown to interfere with Cu metabolism. Table 1 shows a summary of individual mineral functions, deficiencies, toxicities and interrelationships.

Table 1. Summary of individual mineral functions, deficiencies, toxicities and interrelationships

Mineral	Main Functions	Deficiency Symptoms	Major Interrelationships
Calcium (Ca)	Blood coagulation; bone and teeth creation; nerve function; muscle contraction; milk production; cell permeability	Bone abnormalities - osteomalacia in adults and rickets in young animals	The vitamin D is involved in bone deposition and absorption; reduction of PO_4^{3-} diminish absorption; additional Mg declines absorption, substitutes Ca in bone and rises Ca secretion; the ratio Ca:P would be from 1:1 to 2:1
Potassium (K)	Main cation of intracellular fluid where it is involved in acid-base balance; muscle activity and osmotic pressure	Lethargic disorder with elevation of occurrence of comas and death hypokalemia; diarrhea, expanded abdomen and untidy look, muscle weakness, poor growth and stiffness	Extra K decreases Mg absorption; Mg shortage decreases K retention leading to K shortage
Magnesium (Mg)	Bone formation; in glycolytic system acts as enzyme activator	Loss of equilibrium and trembling; hyperirritability with convulsions and vasodilation	May not be toxic and extra Mg may cause upsets of metabolism of Ca and P
Sodium (Na)	Major cation of extracellular fluid where it is involved in acid-base regulation and osmotic pressure; cell penetrability; conservation of cell irritability in normal muscle	Diminish growth; eye disorders with corneal lacerations; in females delayed sexual maturity and in males infertility	Hypertension. staggering gait, blindness and nervous disorders

Phosphorus (P)	Teeth and bone formation; phosphorylation; ATP bonds; in intracellular fluid PO4 chief anion radical; the PO4 is vital in acid-base balance	Bone irregularities - osteomalacia in adults and rickets in young animals	Renal reabsorption and bone deposition where Vitamin D is involved; excess Mg and Ca origins reduction in absorption; extra P can cause urinary
Sulfur (S)	Function in tissue respiration; abundant in the keratin-rich appendages – hoof, hair; component of the vitamins biotin and thiamine	Reduction of growth	It is not toxic
Copper (Cu)	Involved as cofactor in enzyme systems; ; bone formation; hemoglobin synthesis; hair pigmentation and maintenance of myelin in nerve system	Nervous symptoms or ataxia; fading hair coat; swelling of joints and fragility of bones; lameness and anemia	Excess Mo, Zn constrain its use and loading; is toxic at levels 250 ppm with similar symptoms than deficiency
Iron (Fe)	Part of hemoglobin, cytochromes, myoglobin in cellular respiration	Fewer red cells; less than normal amount of hemoglobin and respiratory failure	Cu required for proper metabolism; Ca-P ratio affects absorption; pyridoxine deficiency decreases absorption
Manganese (Mn)	Supposed to be an activator of enzyme systems, fatty acid synthesis and cholesterol metabolism; bone formation; growth and reproduction; amino acid metabolism	Reduced growth; reduced long bones; in males testicular degeneration, in females defective ovulation; damage immunity and brain	Extra P and Ca reduces absorption; not toxic
Zinc (Zn)	Acts as cofactor of several enzyme systems including carbonic anhydrase and peptidases; required for bone tissue growth	Deprived hair growth; thickened skin and rough	Elevated Ca phytate bonds up Zn; extra Zn affects the metabolism of Cu and can origin anemia

Minerals in litterfall

In a study to evaluate and to compare in space (two sites) and time (12 months) the mineral profile of leaf litter collected in the semiarid regions of northeastern Mexico. It was reported that the concentrations of Ca, K, Mg and P in leaf litter were different between sites and among months of collection. The Ca varied from 23 to 43; K from 3 to 14; Mg from 2 to 4 and P from 0.3 to 1.1 g/kg. Except P, all samples had enough macrominerals to satisfy the metabolic requirements of adult small ruminants. It has been determined that P is limited nutrient in the rangelands of northeastern Mexico and Texas, USA; Thus, the normal bone development of small ruminants consuming the leaf litter is limited because the lack of P.

The concentrations of Cu, Fe, Mn and Zn were also different between sites and among months. The Cu content was in a range of 3 to 6; Fe in 103 to 654; Mn in 22 to 42 and Zn in 12 to 25 mg/kg. The Cu in both sites, Mn in site 2 and Zn in both sites were marginal deficient to fulfill the needs of adult range small ruminants. Thus, it is advisable to supplement range ruminants with Cu, MN and Zn when they are consuming leaf litter. Conversely, Fe content, in both sites and all months was sufficient to meet the requirements of range ruminants ingesting leaf litter.

Minerals in foliage from trees and shrubs

Calcium
In a study carried out in rangelands of northeastern Mexico, it has been reported that in general, during summer and winter most plants had higher Ca contents than in other seasons. Despite seasonal variations, all plants had Ca levels that could meet the adult goat requirements (1.3±3.3 g/kg DM in the diet). It seems that foliage from browse plants that grow in semiarid and tropical regions have enough Ca for optimal livestock and wildlife performance. High pH in the soils of these regions may be the cause why shrubs are high in Ca content. It has been stipulated that an average of 5 g/kg of Ca in the aerial parts of higher plants are considered to be adequate. Most plants evaluated in the present book had Ca content above 20 g/kg.

Phosphorous

It has been found that leaves from shrubs collected in a rangeland from Mexico had P concentrations that were significantly different between seasons. During spring, most plants had the highest levels of P and during summer the lowest. All plants had enough P to sustain maintenance needs of range ruminants. Low P and high Ca concentration resulted in an unusually wide Ca:P ratios (from 4:1 to 42:1). Similar wide Ca:P ratios had been reported. However, it seems that browsing small ruminants (goats, sheep and white-tailed deer) can sustain these high Ca:P ratios without being affect P metabolism. In a study to evaluate browse nutrition in rangelands of northeastern Mexico it was reported that P deficiencies in browse plants are mainly associated to the low mineral concentration in soils.

Magnesium

During summer, most plants had higher Mg concentrations than in other seasons. Magnesium requirements for goats are 0.8–2.5 g/kg in their diets. It seems that all tested shrubs of northeastern Mexico and southern Texas, USA, in all seasons, had marginal Mg concentrations to meet adult range goat requirements. Moreover, it has been reported high Mg contents (1.1–8.0 g/kg) in 18 shrubs that growing in Texas, USA. In addition, in some commonly used tropical legumes had Mg concentrations that meet ruminant requirements. Moreover, other studies have found that diets, from esophageal samples by range goats growing in north Mexico, or browse plants from northeastern and northwestern Mexico had sufficient amounts of Mg to meet requirements of adult range goats.

Sodium and potassium

It has been reported that Na content in leaves of shrubs from northeastern Mexico was significantly different between seasons. It appears that all shrubs can be considered as Na non-accumulators because they contain less than 2 g Na/kg DM. Moreover, during all seasons all plants had lower Na content to meet the needs of growing adult range ruminants. High K content in evaluated shrubs in northeastern Mexico and southern Texas, USA could reduce Na absorption of range small ruminants feeding these shrubs because it has been reported that elevated dietary K may decrease ruminal concentration and absorption of Na in sheep and steers. It has been also reported that shrubs had K concentrations significantly different among seasons. During summer, most plants had higher K than in other seasons. Seasonal variation in K content might be related to water availability, because K absorption by the root is linked to the soil moisture.

In a study, during summer, when K was higher, rainfall was also higher (52.9 mm), It seems that a goat weighing about 50.0 kg BW consuming 2.0 kg/day DM of shrubs growing in northeastern Mexico, could eat substantial amounts of K to meet their requirements of K in all seasons. Similar findings were reported by in other studies that evaluated K content in browse species growing in arid and semi-arid regions of the world. High K content might have reduced Na absorption of sheep feeding these shrubs, because it has been reported that elevated dietary K may decrease ruminal concentration and absorption of Na in sheep and steers. However, Na deficiency can be alleviated by supplementing ruminants with common salt.

Copper
All browse plants collected in northeastern Mexico had Cu levels that were significantly different among seasons. Apparently, only during summer, most plants had the higher Cu concentrations than in other seasons. Low Cu concentrations are also reported in shrubs from semi-arid regions and in tropical legumes. Low Cu levels in plants might be caused because the high pH of the soils of these regions which are about from 7.5 to 8.5. In addition, high dietary fiber during dry seasons (winter and spring) might have also reduced availability of Cu.

Iron
It appeared that, all shrubs growing in northeastern Mexico and southern Texas, USA had Fe concentrations that were significantly different among seasons. In addition, all shrubs, in all seasons, contained Fe levels in substantial amounts to meet adult small ruminant requirements. Similar findings were reported in other studies that evaluated the Fe contained in shrubs that grow in semi-arid regions of Mexico. They sustained that Mexican browse species had Fe levels in substantial amounts to meet the Fe requirements of adult range Spanish goats.

Manganese

It seems that all shrubs from northeastern Mexico had Mn concentrations significantly different among seasons. It appears that during summer, most plants had levels of Mn higher than in other seasons. It also appears that all shrubs in all seasons had Mn marginal concentrations to meet the requirements of adult range Spanish goats. Moreover, it has been reported that all evaluated browse plants from south Texas, USA, had low levels of Mn to meet the requirements of grazing cattle (20-40 mg/kg DM). Low Mn levels in shrubs growing in different shrublands of Mexico were also reported. Additionally, it has been found that high concentrations of Ca in Mexican shrubs may increase Mn requirements; possible due to interference with Mn absorption, and availability of Mn may be compromised when high proportion is located in the cell wall.

Zinc

In a study that evaluated that Zn content of all plants collected in rangeland of northeastern Mexico, showed that Zn concentrations were significantly different between seasons. Peaks of Zn levels appeared to be related to summer rainfall. In this study, only a few plants s had sufficient levels of Zn to meet adult small ruminant requirements. Some shrubs that occur in Texas, USA and northeastern Mexico had Zn levels that varied seasonally, but only a few of them had levels of Zn to meet domestic ruminants (goats, sheep and cattle) and white-tailed deer requirements. Conversely, in other study, indicated a relevant potential mineral intake of Zn by range Spanish goats browsing in north and northwestern regions of Mexico, respectively.

Potential mineral deficiencies

In a study that reviewed the possible mineral deficiencies of native browse plants in northeastern Mexico, described that seasonal changes in dry and wet conditions seem to have had larger effect on sheep choosing their diets than on goats. The alterations were also revealed in different stages of monthly mineral consumptions, as reveled from extrusa samples of esophageal fistulated sheep. Ca, Mg, Na, Cu, and Fe concentrations were plentiful lesser in the sheep diets than in those of goats, meanwhile K, Mn and Zn were much higher. However, Mn, Cu and Ca were scarce for sheep needs in some months, and Mg and K during the year. Thus, the need for mineral supplementation for sheep is also recommended with the purpose to avoid problems in health and performance because of deficiency problems.

White-tailed deer, by contrast in the similar region as the goats and sheep, selected diets similar to goats from 94% browse constantly during the year, 5% forbs and 0.6% grasses, as determined by micro histological examines of the white-tailed deer feces, and with *Vachellia rigidula* and *Parkinsonia texana* browse species controlling in the rangeland. Mineral concentrations of the diets seemed to be selected by the white-tailed deer were not as varied monthly as in the sheep or goat diets, and appeared to be satisfactory for assessed needs taken from beef cattle requirements in the lack of specific standards for white-tailed deer. However, during six months of the year, in summer and fall, the provisions of Zn were under the range of beef cattle needs, and supplementation with this element to deer seems to be suitable.

Additional studies recognized the mineral concentrations of browse and native grass species that were usually ingested by dual-purpose goats in selection of their diets in the northeastern Mexico. Some species were prominent in mineral concentrations over other browse species. In contrast, the accessible browse species and their mineral concentrations in northwestern area of Mexico were also recognized. These findings may help relieve seasonal mineral deficiencies in range ruminants. The species may provide in a less expensive and more regular mineral complementation than would profitable supplements, although none of these browses species had sufficient minimum contents of Mg, Cu, Mn, and Zn below the needed different absorption situations.

However, it is stimulating that some browse species seem to be prominent in some elements and during some seasons of the year. Therefore, they may deserve favorable attention in range management, such as *Celtis ehrenbergiana* that actually is presently less frequent in the total forage cover in northeastern Mexico, and appear to be good sources of Ca and Mg. *Vachellia farnesiana* seems to have relatively high content during some seasons of the frequently deficient microminerals Cu, Mn and Zn. In other studies, *Opuntia*, *Vachellia*, and *Quercus* genus were recognized as potentially beneficial suppliers of higher levels of macro and microminerals to browsing small ruminants. It would be remarkable to determine, whether and how these browse plants react to mineral fertilizations of the range. Prominent browse in arid northwestern Mexico were *Vachellia rigidula*, *Prosopis juliflora* and *Parkinsonia texana* browse plants, as well as some other legumes and nonlegumes.

CHAPTER 3

Plant mechanisms against herbivory

Introduction

Foliage from trees and shrubs have represented an important source of protein for grazing ruminants. However, in some instances, not only has their CP digestibility been detected to be short, but also in several circumstances of livestock death have been associated with high tannin content of some forages. Every bite an herbivore takes comes at the expense of a plant. Some plants have evolved tolerance to herbivory, growing compensatory tissues so rapidly that reproduction can sometimes increase with light damage. By comparison, other plants have evolved traits that reduce consumption by herbivores, or resistance. Due to herbivores depend on plants for food; natural selection helps herbivores that overcome plant confrontation, thus prompting plants to evolve new forms of resistance. This evolutionary competition between herbivores and plants has occasioned in an extensive selection of opposition characters in terrestrial plants including diminished apparency to herbivores and mechanical, chemical, and indirect defenses. Moreover, plants evade herbivory by hiding, building mechanical defenses, creating and obtaining chemical toxins, and employing predatory. Thus, plants are not the abandoned targets of herbivory but protect themselves against the loss of properties and energy, permitting for better investment in reproduction and existence.

Their first line of defense, the plant surface

The cuticle and the periderm, as well delaying water loss, offer inert fences to microorganism entrance. A varied collection of vegetable complexes, usually denoted to as secondary metabolites, also protects plants against a diversity of herbivores and pathological microorganisms.

Mechanical defenses

In several tree and shrub species, thorns and spines are common and may affect browsing capacity by reducing bite mass and diminishing biting and chewing rates. They make it problematic for animals to bit leaves off stems, which conduce animals to select individual leaves. Spines also slow chewing rate by necessitating herbivores to cautiously operate plants in their mouths to evade pain and damage. However, the effect of spines and thorns on ingestion depends on the mouth size of the foraging animal. Most browsing small ruminants have lips and tongues that are very agile and can more easily select leaves and avoid thorns. Goats, for instance, with their mobile and narrow muzzle, can manipulate their mouths more easily among thorns to pluck small leaves, such that thorns may be less effective in reducing cropping rates. This partially may explain why sheep are less effective browsers than goats.

Plant morphology, which is affected by browsing, may also, influences browsing rates and feed intake. However, plants whose leaves grow on old shoots incline to result in high bite rates and reduced feed intake rates, plants with leaves that grow on young edible shoots allow larger bite sizes and relatively higher intake rates. Because bite size plays a major role in influencing intake rates, plants and plant parts that allow animals to have lager bite sizes are mostly selected. Animals prefer plants from that they may harvest bigger bite sizes, which increase instantaneous intake rates and daily feed intake. Nevertheless, instantaneous intake rates might not explain longer-term daily food intake because they vary with the hunger status of the animal and with forage availability. As animals satiate on particular food items, they are more likely to select a different food item not only on the basis of the amount of biomass an animal can harvest but due to a need to meet other nutritional requirements or to avoid excesses of nutrients and toxins.

Other factors such as plant phenology may affect accessibility of leaf material for small ruminants during the dry season and reduce intakes rates and daily feed intake. Plants that loose their leaves during the dry season (deciduous plants) provide less forage material during these periods compared to perennial species whose browsed material is available during the year. Small ruminants foraging in shrublands combining herbaceous and browse material may select feed particles from both green and dead plant material. Moreover, the diverse forage situation of shrubby rangelands suggest circumstances that may favor a very fast intake for small ruminants that differentiate between forage materials while at the same time keeping an intake rate higher than that observed during their meal.

Plant secondary metabolites

Secondary metabolites are divided into three major groups Plant secondary metabolites can be divided into three chemically distinct groups: terpenoides (aromatic oils, resins, waxes, steroids, rubber, carotenoids), nitrogen-containing compounds (often toxic, e.g., strychnine, nicotine, caffeine, cocaine, capsaicin) and phenolics (flavonoids, tannins, lignin, salicylic acid). These chemicals are extremely diverse; many thousands have been identified in several major classes. Each plant family, genus, and species produces a characteristic mix of these chemicals, and they can sometimes be used as taxonomic characters in classifying plants. Humans use some of these compounds as medicines, flavorings, or recreational drugs.

The apparent lack of primary function in the plant, combined with the observation that many secondary metabolites have specific negative impacts on other organisms such as herbivores and pathogens, leads to the hypothesis that they have evolved because of their protective value. Many secondary metabolites are toxic or repellant to herbivores and microbes and help defend plants producing them. Production increases when a plant is attacked by herbivores or pathogens. Some compounds are released into the air when insects attack plants; these compounds attract parasites and predators that kill the herbivores. Recent research is identifying more and more primary roles for these chemicals in plants as signals, antioxidants, and other functions, so "secondary" may not be an accurate description in the future.

Consuming some secondary metabolites can have severe consequences. Alkaloids can block ion channels, inhibit enzymes, or interfere with neurotransmission, producing hallucinations, loss of coordination, convulsions, vomiting, and death. Some phenolics interfere with digestion, slow growth, block enzyme activity and cell division, or just taste awful.

Most herbivores and plant pathogens possess mechanisms that ameliorate the impacts of plant metabolites, leading to evolutionary associations between particular groups of pests and plants. Some herbivores (for example, the monarch butterfly) can store (sequester) plant toxins and gain protection against their enemies. Secondary metabolites may also inhibit the growth of competitor plants (allelopathy). Pigments (such as terpenoides, carotenes, phenolics, and flavonoids) color flowers and, together with terpene and phenolic odors, attract pollinators.

Terpenes

The terpenes, or terpenoids, create the major type of secondary metabolites. Most of the varied compounds of this type are water insoluble. They are created from acetyl-CoA or its glycolytic groups. Studies conducted *in vitro* had suggested that terpenes found in *Artemisia tridentata* eliminate microbes in the digestive system, thus reducing foliage digestion. Terpenes are volatile organic compounds that are ejected quickly, and thus have little or not effect on rumen microbes and digestibility. However, it has been found measureable small amounts of terpenes in the stomach of marsupials signifying that terpenes may affect rumen microorganisms. Pharmacological studies on sheep reveled significant amounts of monoterpenes are absorbed from the rumen and have considerable influence on the feeding behavior of ruminants. A strong estimation of the effects of terpenes is still necessary to advance knowledge that will form a basis for mitigating limiting effects of terpenes to herbivory.

Terpenes such as monoterpene esters called pyrethroids, localized in leaf and flower of Chrysanthemum species, show outstanding insecticidal action. Both natural and synthetic pyrethroids are current elements in commercial insecticides due to their little perseverance in the atmosphere and their insignificant poisonousness to mammals. Many plants have combinations of volatile monoterpenes and sesquiterpenes, named essential oils that give a typical scent to their vegetation. Peppermint, lemon, basil, and sage are cases of species that have essential oils.

Nitrogen-containing compounds

Many plant secondary compounds have N in their chemical structure. Comprised in this group are those defenses against herbivory as alkaloids and cyanogenic glycosides that are of significant attention due to their poisonousness to humans and their therapeutic properties. Many N secondary compounds are created from mutual amino acids. Examples of nitrogen-containing compounds are alkaloids, cyanogenic glycosides, glucosinolates, and nonprotein amino acids.

Many alkaloids are now supposed to role as defenses in contradiction of herbivores, particularly mammals, due to plant defensive compounds and even use them in their particular protection. Cyanogenic glycosides release the toxic gas hydrogen cyanide (HCN). The break of cyanogenic glycosides in plant species is an enzymatic procedure. Plants that elaborate cyanogenic glycosides similarly make the enzymes required to hydrolyze the carbohydrate and release HCN. The glucosinolates, or mustard oil glycosides, break down to liberate protective compounds. Found mainly in the plants of the family Brassicaceae. The glucosinolates break to produce the substances accountable for the smell and taste of plants like radishes, broccoli and cabbage,

Tannins

Tannins are polyphenolic compounds that are broadly categorized into two major groups:

1. Condensed tannins, or proanthocyanidins, consisting of oligomers of two or more flavan-3-ols, such as catechin, epicatechin, or the corresponding gallocatechin.

2. Hydrolysable tannins, consisting of a central core of carbohydrate to which phenolic carboxylic acids are bound by ester linkage.

Tannins have a very high affinity for proteins and form protein-tannin complexes. The ingestion of a plant containing condensed tannins decreases nutrient utilization, protein being affected to a great extent, and decreases feed intake. On the other hand, hydrolysable tannins are potentially toxic to animals. Consumption of feeds containing high levels of hydrolysable tannins cause liver and kidney toxicity and lead to death of animals. Oak and yellow wood poisonings are attributed to hydrolysable.

The tannins are widely distributed throughout the plant kingdom, especially among trees, shrubs and herbaceous leguminous plants. The great variety of species over which these compounds are found has grown as detection techniques have improved. Tannins are high molecular weight, water-soluble polyphenols that form reversible complexes with proteins through pH-dependent hydrogen bonding and hydrophobic interactions. Hydrolysable tannins (HT) contain a carbohydrate core esterified with gallic or hexa-hydroxyl-diphenic acids. Binding of HT to abomasal mucosal proteins causes lesions that result in diarrhea or constipation. Hydrolytic products of HT are absorbed from the small intestine; disrupt liver and kidney function, and cause photosensitization and dehydration. Condensed tannins (CT) or proanthocyanidins are oligomers of flavan-3-ols or flavan-3,4-diols that are linked by C-4/C-8 or C-4/C-6 inter flavan bonds.

Variations in chemical reactions of CT arise from differences in monomeric constituents; inter flavan bond type, polymer length and branching, molecular weight, and concentration. High concentrations of CT in ruminant diets result in formation of stable, insoluble complexes with digestive enzymes and proteins in feed, saliva, and microbial cells, decreasing feed intake and digestibility and increasing fecal N excretion. Complexation of metal ions by CT can result in microbial mineral deficiencies. Protection against the negative effects of CT is provided by proline-rich salivary proteins in some ruminants and can be achieved with polyethylene glycol supplements, which disrupt tannin-protein complexes.

Tannins in ruminant nutrition

The tannins in some herbivores reduce the palatability and digestibility of dry matter and protein. Sometimes they act as toxins, rather than as inhibitors of digestion. The diversity of effects of tannins on digestion is partly due to the physiological capacity of animals to use and partly due to differences in the chemical reactions of different types of tannins. In some mammals, salivary proteins react with the tannins. In white tailed deer this tannin-binding salivary proteins are glycoproteins containing large amounts of proline, glycine and glutamate/glutamine.

Tannins are an important part of the characteristics that determine the palatability of plants by herbivores due to the astringent properties of these compounds. By this mean plant reduces herbivory frequency by ruminants and improve their chances of survival. It has been found that plants that receive more herbivore attack are able to increase their concentration of tannins. In general, tannins are more abundant in the parts of the plant that are more likely to be eaten by herbivores. Numerous reports illustrate the effects of environmental and seasonal factors as well as of phonological development. Very briefly, high temperatures, water stress, extreme light intensities and poor soil quality increase the tannin content of plants.

Tannins can be beneficial or detrimental to ruminants, depending on which (and how much) is consumed, the compound's structure and molecular weight, and on the physiology of the consuming species. It appears that the consumption of plant species with high CT contents (generally above 4.0% of dry matter, DM) animal´s voluntary feed intake is reduced, while medium or low consumption (below 5% DM) seems not to affect it. Three main mechanisms have been suggested to explain the negative effects of high tannin concentrations on voluntary feed intake:

1. Reduction in feed palatability. A reduction in palatability could be caused through a reaction between the tannins and the salivary mucoproteins, or through a direct reaction with the taste receptors, causing an astringent sensation.
2. Slowing of digestion. Slowing the digestion of dry matter in the rumen impairs the emptying of the digestive tract, generating

signals that the animal is full and providing feedback to the nerve centers involved in intake control. Thus, this could influence voluntary feed intake more than a reduction of palatability

3. Development of conditioned aversions. The third mechanism is based on the identification of negative post-prandial consequences following the ingestion of tannins, and the subsequent development of conditioned aversions. The microorganisms of the rumen play a fundamental role in the nutrition of ruminants. It would therefore seem probable that the post-prandial consequences of ingesting tannin-rich feeds are mediated by factors relating to microbial fermentation, by reducing de number of cellulolytic and proteolytic bacteria.

Although rumen microorganisms have metabolic mechanisms that prevent, tolerate and regulate the consumption of CT, the negative effects of animal productivity are attributed to establishment of CT with cellulose and protein forming complexes resistant to ruminal degradation, which reduce 1) adherence of microorganisms to substrates, 2) microbial colonization, 3) the activity of the enzyme 1,4-endoglucanase, 4) degradation of protein, amino acid absorption in the intestine, 5) digestibility and 6) consumption.

Negative effects of tannins

Tannins performance as a protection machinery in plants in contradiction of pathogens, herbivores and antagonistic environmental situations. In general, tannins produce a negative reaction when ingested. These aspects may be immediate such as astringency or a nasty or disagreeable sensitivity or might have a deferred answer linked to antinutritional/ poisonous properties. Tannins negatively disturb feed consumption, feed digestion, and effectiveness of productivity. These properties differ conditional on the concentration and kind of tannin consumed and on the acceptance of the animal that in sequence is reliant on on features such as kind of gastrointestinal tract, feeding behavior, animal size, and decontamination mechanisms. The places of action of tannins are 1) oral cavity (mastication breaks the plant cell walls and exposes carbohydrates and proteins to tannins), and 2) rumen and gastrointestinal tract lumen (unbound tannins complex proteins from the diet and metabolic proteins, for example microorganisms, enzymes, cells from the epithelium).

Positive effects of tannins

At low concentrations (less than 5%), CT decrease proteolysis of dietary proteins, rumen ammonia production, and rumen bacterial biomass, and increase N flow to the abomasum and absorption of essential amino acids in the small intestine. Results include increased animal weight gain; improved fiber, meat, and milk production; and higher ovulation rate. Protein-CT complex formation also reduces rumen gas formation and prevents production of stable foam in the rumen, alleviating bloat in ruminants consuming protein-rich diets. Anthelmintic properties of CT in ruminants are associated with improved nutrient supply to the lower gastrointestinal (GI) tract. Positive effects of CT on GI nematode parasites include lower fecal egg counts, decreased worm burdens, and inhibition of egg hatch and larval development.

Forages containing high tannin content are non-bloating due to tannins bind additional plant dietary proteins, precipitating them out of rumen fluid, and in the procedure, avoiding the formation of the stable foam that is distinguishing of meadow bloat. Two general features of tannins important to browsing ruminants are the inhibition of bloat and the elimination of internal parasites. The bloat happens when a considerable quantity of fresh, rich in protein foliage, such as alfalfa, is digested rapidly, causing in a fast growth in the protein concentration of the rumen. This may origin the rate of microbial fermentation in the rumen to growth, and consequences in quick addition of CO_2 and CH_4 gases in the rumen. Microbial mucus, cellular membranes of plants and proteins all conglomerate with fermentation gases to make a steady foam that is observed as a liquid at the valve conducing from the rumen into the esophagus, producing it to persist locked. As the gases confined in the rumen continue to store, the rumen converts inflated, interfering with breathing and blood movement. Without treatment, bloat may effect in decease from asphyxia or cardiac detention.

The destruction of internal parasites by tannins, precisely the elimination of several nematode species, has been recognized for purified tannins from browse species used as dietary complements. However, the tannin content the chemical arrangement and the type of nematode determine the effect of tannins on nematodes. The efficiency of tannins also varies by the period of development of the nematode, and the site in the gastrointestinal tract where the tannin is effective.

Effect of polyethylene glycol on tannins

Polyethylene glycol (PEG), a non-nutritive synthetic polymer, has a great attraction for phenolic compounds, especially tannins, and thus neutralizes them by creating tannin-PEG compound. Therefore, PEG can avoid their creation or release protein from tannin-protein compounds, and it has been utilized to alleviate hostile effects of secondary compounds on rumen fermentation, as well as increase performance (milk yield and growth) of ruminants fed diets containing high secondary compounds. For instance, addition of PEG to browse and herbaceous legumes high in secondary compounds increased *in vitro* gas production and short chain fatty acid production, even though microbial N production and efficacy of microbial protein synthesis was diminished. In other study, addition of PEG increased *in vitro* fermentation, digestibility and metabolizable energy, while diminished productivity of microbial protein synthesis by retreating effects of their secondary compounds. Meanwhile, the extent of benefits of PEG addition diverse to some extent by browse leaf, with *Eucalyptus camaldulensis* and *Schinus molle* having the main total development, *Cassia fistula* the lowest and *Chorisia speciaosa* intermediate, it did not impact their total nutritive position.

Effect of wood ashes on tannins

The high alkalinity of wood ash solution succeeded in reducing tannins level in *Acacia cyanophylla* leaves. The extent of this effect depends on the level and source of wood ash and on treatment duration. Wood ash treatment could be emphasized to deactivate tannins but provision of an adequate amount of energy would be necessary if better utilization of tannins-rich shrubs like *A. cyanophylla* is targeted. Soaking acacia foliage for 1 day in a solution containing 180 g/kg of wood ash resulted in optimum tannin deactivation. This solution could be used up to five times and then discarded. Further research is required to explore the combined effect of wood ash treatment and energy supply on the nutritive value of tanniniferous shrubs, and consequently on animal performance.

CHAPTER 4

Leaf litter as a feed resource

Introduction

Litterfall is dead plant material that has fallen to the ground. It is also called plant litter, leaf litter, tree litter, soil litter, or duff, and is composed by leaves, bark, needles, and twigs. Litterfall and litter decomposition are key processes in nutrient cycling of forest ecosystems. Litterfall plays a fundamental role in nutrient turnover and in the transfer of energy between plants and soil, the main source of organic material and nutrients being accumulated in the highest layer of the soil. Nutrient release from decomposing litter is an important internal pathway for nutrient flux in plantation ecosystems. Evaluation of litterfall production is important for understanding nutrient cycling, forest growth, successional pathways, carbon fluxes, disturbance ecology, and interactions with environmental variables in forest ecosystems. However, litterfall inputs vary widely among forest ecosystems in terms of quality and quantity. The quality of soil organic matter is of great importance for the majority of the functional processes occurring in the soil of forest ecosystems. The mixture of leaves represents the main component of the litter fall that remains productive throughout the year, especially in the dry and winter months. Research is needed concerning the nutritional quality and fermentative profile of the mixture of litter fall leaves when are fed to range ruminants.

Nutritional value of leaf litter

Litterfall may signify an excellent food source for small ruminants that continue dynamic during all year, particularly in drought and winter seasons. In an investigation, it was found that the diet sika deer (*Cervus nipponyakushimae*) on Nakanoshima Island, Japan, contained of fallen leaves of browse and, about 75% of the annual dietary plant components comprised of forest litter, although deer had admittance to plentiful living palatable leaf materials in the experimental area. Thus, sika deer in several circumstances role ecologically as decomposers instead than main consumers. In other research, it was determined that leaf litter, in unharvest locations of wintering north areas of USA, was high compared to the accessibility of more normally measured woody plants (understory shrubs and trees), and can be more edible and of more nutritional quality for white-tailed deer.

Browse species is the basic feed resource in winter season for most range small ruminants of northeastern rangelands, particularly deer and goats, and is frequently considered a restrictive source. The theory of equilibrium expects that browsing will diminish the sources accessible down to a serious threshold where the degree of growth of browse plants and small ruminant populations ranges to zero. Even though of these active variations, high fertility, time lags in the degree of increment of ruminant animals compared to the decay in feed resources, and a great neonatal. In a study, it was determined that tree lichens were more digested than browse plants. Leaf litter and lichens dropped in winter season can account for about 85.0% of the fodder accessible to deer in unharvested woodland locations in Maine, USA. The tree lichens accounted for 6.0% of the total leaf litter vegetation but contributed about 31.0% of the accessible energy for wintering the deer diet. Therefore, it was found that great number of sika deer (*Cervus nippon*) were preserved by leaf litter in deciduous woodlands. The richness of fallen leaves from mature plants is not related to browsing in a long time basis that announces a chronological uncoupling among the effect of deer selectivity on balsam fir seedlings and the undesirable response from enrolment disappointment of mature balsam fir on the deer numbers. This elucidates that the structure is vulnerable to be obligatory into another management.

Intake of fallen leaves by range ruminants has been reported for white-tailed deer, mule deer (*Odocoileus hemionus*), elk (*Cervus elaphus*), black-tailed deer and range goats. In a study, it was found that different tendencies of habitat use and the dietary content, provided extra funding for the hypothesis that fallen leaves from balsam fir gives an substitute feed resource supporting high deer numbers on Anticosti Island. The same happened in mixed rangelands by wintering white-tailed deer in southern New Brunswick. In other study, it was determined that leaf litter in unharvested locations was high compared to the accessibility of more usually measured woody plants (understory shrubs and trees), and can be more edible and of higher nutritional quality. Nonetheless, accessibility of woody plants in winter season can be more expected than leaf litter.

In winter season, ruminants in boreal woodlands have to manage with high energetic costs linked to movement in deep snow and compact fodder plenty and nutritional value. At high numbers, ruminants face extra restrictions, due to substantial browsing diminishes accessibility of browse plants, the chief resource of fodder in winter season. Under these negative situations, large ruminants might forage on another feed resources such as autonomous of browsing density, like fallen leaves or windblown plants. It was determined that to recompense for the lack and random feed provisions, deer choose habitat classes, but mainly locations within those habitat classes, where the possibility of discovery browse plants, leaf litter, and windblown plants was utmost.

In other research under taken during the winter season, white-tailed deer responded to another feedstuff resources that converted accessible dependent on stochastic events, like tough winds, dry or abundant dense snow situations. Due to they were tested with the irregularity of food accessibility, deer obtained their habitat to utilize more intensively regions of their home range where the prospect of discovery fallen leaves and windblown plants was better. Deer in addition accustomed their winter movement and actions to reserve energy when environmental situations were the best. When favored feed sources were reduced during winter, deer designated varied habitat varieties or regions where plant multiplicity was high. It was established the necessity of deer on the plant *Abies balsamea* in winter season and offered extra nutrition to the hypothesis that leaf litter from *A. balsamea* provided another feed resource associated to high deer numbers in a boreal forest.

In a study carried out with arboreal lichen production and availability for Rocky Mountain elk (*Cervus elaphus nelsoni*), white-tailed deer (*Odocoileus virginianus*), and mule deer (*O. hemionus*) during winter in western Montana, USA. It was collected lichen litterfall inside and outside exclosures to assess deer and elk use of lichens and recorded associated tree stand characteristics. *Bryoria* spp. and *Nodobryoria* spp. composed >99% of the lichen litterfall. Lichen litterfall use by deer and elk on 2-yr sites averaged 7.91 kg/ha for the severe winter, and 6.02 kg/ha for the milder winter. The greater use of lichens in winter was probably due to increased ungulate densities in forested habitats during a severe winter.

Because there is not exist work done on nutritional assessment of the various fractions of litter deciduous tropical forest northeastern Guarico state of Venezuela; consequently, it was aimed at evaluating the nutritional composition of the various litter fractions (fallen leaves, peduncles, shafts, dry fruits, fleshy fruits and small particles difficult to identify). It was found that production of dry matter was 1061.9 g/m^2/year and the litter in a tropical deciduous forest represented a valuable resource with nutritional potential to supplement ruminant animals during the dry season.

In other study carried out in Michoacán, Mexico to determine the chemical composition of the leaf litter composed by species of deciduous forest that are consumed by livestock in the dry season. It was reported that fallen leaves exhibited a great variation in chemical composition. It represented an important source of food for livestock during the dry season. The average value of protein was 10.0% with variations between 4.9% and 22.5%. The 40% of species were preferred, 33 representing 82.5% had values higher than 8% of protein, one of the requirements to be consider as fodder. The FDN presented an average of 39.2% with fluctuations between 22.6% and 55.1%. The 30% of the species were between 20% and 35%, suggesting good degradability for this group and acceptable to the other species. The native species represent a significant potential as a source of feed for range ruminants in the dry season so as litter and as grazing in the rainy season. In addition to its effect on the water cycle, soil protection, wildlife protection, production of medicinal plants, etc. While litter is a source of food, there must be a balance in productivity in order to regulate and streamline the biogeochemical cycles with coefficients of rangeland.

Nutritional characteristics of leaf litter from the Thornscrub vegetation

The Tamaulipan Thornscrub of the semiarid plains of Northeastern Mexico contains a wide diversity of shrub species that have developed morphological and physiological characteristics to adapt to adverse climatic factors; in addition, could be important animal feed resources in times of scarce feed availability to range ruminants (goats and sheep) and wildlife (white-tailed deer). The mixture of leaves in this ecological and biological type of scrub represents the main component of the litter fall that remains productive throughout the year, especially in the dry and winter months. A research was conducted to evaluate the nutritional quality and fermentative profile of the mixture of litter fall leaves from shrubby native species in the Tamaulipan Thornscrub of northeastern Mexico. It was hypothesized that the mixture of leaf fall from the Tamaulipan Thornscrub from northeastern Mexico, is a good source of nutrients to meet the metabolic requirements of range small ruminants during dry and winter months. Thus, the objectives of the study were to evaluate during 12 consecutive months, the nutritional quality of samples of litter fallen leaves and their fermentation characteristics, treated or not with PEG, in two sites located in the state of Nuevo Leon, Mexico.

Litter fall samples (leaves, reproductive structures, twigs and miscellaneous residues) were collected at 15-day intervals than were grouped into only one in each month for chemical and digestion analysis. The metabolizable energy (ME) content was calculated as ME (MJ/kg DM) = 2.20 + 0.136 GP24$_h$ + 0.057 CP + 0.0029 EE^2. Where: GP24$_h$ is gas production after 24 h of incubation (ml gas/200 mg DM); CP is the crude protein (g/kg DM); and EE is the ether extract (g/kg DM). The chemical content of samples varied between sites and among months (Tables 4.1 and 4.2). The in vitro gas production, ME and microbial protein (MP) content of samples in absence or presence of PEG, were not significantly different between sites, but significant variations were registered among months (Table 4.3). Samples treated with PEG had significantly higher in vitro gas production than samples without PEG.

A similar response as in vitro gas production was observed for ME and MP (Table 4.4).

Table 4.1. Chemical composition (%, dry matter) of leaf litter fall samples from the Tamaulipan Thornscrub in the state of Nuevo Leon, Mexico

Sites	Months	Ashes	NDF (om)	ADF (om)	Lignin	Cellulose	Hemicellulose
Site 1	Jan	10.2	40.4	28.5	20.0	8.5	11.9
	Feb	11.9	44.4	32.0	25.0	7.0	12.4
	Mar	5.8	45.4	33.1	24.8	8.3	12.3
	Apr	11.3	47.5	36.3	22.8	13.5	11.2
	May	4.2	46.6	35.3	21.7	13.6	11.3
	Jun	9.8	45.6	32.2	21.0	11.2	13.4
	Jul	7.9	37.2	20.1	11.8	8.3	17.2
	Aug	9.3	36.8	26.7	12.7	14.0	10.1
	Sep	4.6	33.4	22.6	15.6	7.0	10.8
	Oct	5.9	43.4	31.0	23.2	7.8	12.4
	Nov	9.3	42.8	32.0	25.1	6.9	10.8
	Dec	9.4	33.3	21.4	16.1	5.2	12.0
Site 2	Jan	14.1	42.0	29.7	21.8	7.9	12.3
	Feb	16.1	44.6	32.9	25.2	7.7	11.7
	Mar	18.0	42.2	31.3	20.8	10.5	10.9
	Apr	17.2	43.7	32.3	23.5	8.8	11.4
	May	17.1	43.4	33.1	20.7	12.4	10.3
	Jun	16.4	43.5	31.4	22.0	9.5	12.1
	Jul	16.2	40.0	22.9	18.4	4.6	17.0
	Aug	15.2	31.8	21.6	15.8	5.8	10.2
	Sept	16.2	37.1	26.0	19.6	6.4	11.1
	Oct	14.2	43.2	31.1	21.0	10.1	12.1
	Nov	15.1	39.8	28.5	23.0	5.6	11.3
	Dec	15.6	30.3	19.6	13.9	5.6	10.7
Grand mean		12.0	40.8	28.8	20.2	8.6	11.9

NDF(om) = neutral detergent fiber (organic matter)

ADF(om) = acid detergent fiber (organic matter)

In this study, addition of PEG to shrub foliage results in increased ME values; nevertheless, values of ME (10 MJ/kg DM) of fallen leaves without PEG in this study may satisfy the requirements of maintenance of range ewes (8.4 MJ/kg DM) and meat does (8.0 MJ/kg DM) in early pregnancy with a single fetus (NRC, 2007). High ME content of studied samples might be due to its high IVOMD (63%). Moreover, Differences in gas production at 24 h with or without PEG may be a result of observed differences in the ash, lignin, ADF and NDF content in studied samples.

Table 4.2. Means of crude protein (CP, %), condensed tannins (CT, %), ether extract (EE, %) and *in vitro* digestibility of organic matter (IVDOM, %) of leaf litter samples

Sites	Months	CP	CT	EE	IVDOM
Site 1	Jan	11.9	0.9	3.4	64.8
	Feb	11.3	1.0	3.9	65.9
	Mar	8.1	0.6	2.9	60.8
	Apr	10.5	0.4	2.0	55.1
	May	12.8	0.5	2.3	53.6
	Jun	11.5	0.8	2.4	63.9
	Jul	11.3	0.3	2.8	66.3
	Aug	10.6	1.3	2.1	71.3
	Sep	11.4	0.1	2.4	67.7
	Oct	11.0	0.3	1.8	56.1
	Nov	10.9	0.3	2.4	50.5
	Dec	9.2	0.4	2.2	69.1
Site 2	Jan	13.1	0.9	5.3	62.7
	Feb	13.1	0.7	3.5	65.2
	Mar	11.1	0.6	3.4	65.4
	Apr	12.6	0.9	3.6	61.9
	May	12.6	1.0	3.3	61.5
	Jun	11.9	1.0	2.6	59.3
	Jul	12.2	0.6	3.2	67.0
	Aug	12.5	0.1	3.2	75.1
	Sept	11.6	0.1	3.5	69.0
	Oct	11.8	0.5	3.1	60.4

Nov	11.3	1.1	2.3	52.3
Dec	11.2	0.4	2.6	65.4
Grand mean	11.5	0.6	2.9	63.0

Table 4.3. *In vitro* gas production at 24 h incubation time (Pgas 24h, ml/200 mg), metabolizable energy (ME, MJ/kg) and microbial protein (MP, μmol) of leaf litter fall samples treated (+) with or without (-) polyethylene glycol (PEG)

Site	Month	Pgas 24h		ME		MP	
		-PEG	+PEG	-PEG	+PEG	-PEG	+PEG
Site 1	Jan	61.1	69.1	11.2	12.2	9.0	11.9
	Feb	50.3	57.4	9.7	10.2	8.6	9.3
	Mar	57.2	65.3	10.4	11.5	7.9	8.9
	Apr	46.5	51.5	9.2	9.7	14.9	15.4
	May	67.1	74.3	11.9	13.2	15.2	19.0
	Jun	51.0	55.1	9.8	10.3	13.3	18.5
	Jul	58.2	66.2	10.8	11.8	13.9	14.8
	Aug	58.4	63.4	10.5	11.9	11.5	12.4
	Sep	50.2	56.2	9.6	10.1	9.5	12.4
	Oct	47.3	51.0	9.2	10.4	11.5	14.9
	Nov	53.4	56.0	10.0	10.5	10.4	13.2
	Dec	46.5	52.4	9.0	10.6	11.6	13.2
Site 2	Jan	56.1	59.3	10.7	11.0	10.1	11.4
	Feb	42.2	52.4	8.6	10.1	11.5	12.9
	Mar	45.3	56.0	9.0	10.5	10.4	12.6
	Apr	55.4	66.1	10.4	11.8	12.1	14.2
	May	50.0	56.1	9.0	10.6	12.6	13.4
	Jun	56.1	59.0	10.5	10.9	8.4	8.7
	Jul	47.0	53.2	9.3	10.4	12.5	15.7
	Aug	64.2	76.2	11.5	13.3	14.5	19.5
	Sept	47.3	54.1	8.6	8.7	12.7	18.5
	Oct	52.1	57.2	9.4	10.0	10.6	13.1
	Nov	54.2	58.4	9.2	10.8	11.1	13.5
	Dec	55.2	59.5	9.1	10.7	11.4	13.1
Grand mean		53.2	59.1	9.9	10.9	11.5	13.8

Moderate gas production at 24 h (53 ml/200 mg) might reflect the values of IVOMD and ME content, even in fallen leaves without PEG. It was claimed that significant differences among shrub foliage treated with or without PEG, which demonstrates the affinity of PEG to bind tannins. Concentrations of MP in leaf litter samples were higher than those reported previously in foliage of native forbs, which might be explained by a good synchronization of fermented soluble carbohydrate (100 - (NDF + CP + EE + ash) = 33%; NRC, 2001) and CP level (> 11%) during the incubation period in this study.

Table 4.4. Effects of polyethylene glycol (PEG) on *in vitro* gas production at 24 h incubation time (Pgas 24h), metabolizable energy (ME) and microbial protein (MP) of leaf litter samples treated with or without polyethylene glycol (PEG)

Concept	Pgas24h (ml/200 mg)	ME (MJ/kg DM)	MP (µmol)
Samples without PEG	53.2[b]	9.9[b]	11.5[b]
Samples with PEG	59.1[a]	10.9[a]	13.8[a]
Mean	56.1	10.4	12.7
SEM[1]	1.1	0.2	0.3

Data related to important biomass production as well as crude protein, *in vitro* organic matter digestibility, gas production at 24h, microbial protein synthesis and metabolizable energy and low content of cell wall constituents; support the potential of leaf litter for range small ruminants in semiarid regions of northeastern Mexico, mainly in site 1. Knowledge on the nutritional attributes of leaf litter of the Tamaulipan Thornscrub vegetation may become a useful tool for those interested in range small ruminant nutrition practices, with economic benefits with a reduction in the cost of the rations. In addition, for general information of producers or researchers who observe livestock ingesting considerable amounts of fallen browse leaves.

CHAPTER 5

Nutrition of *Caesalpinia mexicana* A. Gray

Introduction

Caesalpinia mexicana is a species of family Fabaceae (Leguminoseae). The common names are Mexican Holdback, and Tabachin del Monte. It is native to south to central Mexico and the extreme lower Rio Grande Valley of Texas in the United States of America. Its origin is northeastern Mexico and southern Texas. In Mexico, it is distributed in Jalisco, Oaxaca, Nuevo Leon, Sinaloa, Tamaulipas and Guerrero. Part of tropical deciduous forests, medium semi-deciduous forests, sub montane scrub and some oaks, between 750 and 1,500 meters. It is an intensely exploited plant. In Nuevo Leon is used medicinally against nuisance. The tender branches are used as fodder for cattle, goats and sheep.

Caesalpinia mexicana

Table 5.1. Taxonomic characteristics of *Caesalpinia mexicana* A. Gray

Kingdom	Plantae – Plants
Sub kingdom	Tracheobionta – Vascular plants
Super division	Spermatophyta – Plants with seeds
Division	Magnoliophyta – Plants with flowers
Class	Magnoliopsida – Dicotyledonous
Subclass	Rosidae –
Order	Fabales –
Family	Fabaceae – Leguminosae
Genus	*Caesalpinia*
Species	*mexicana* A. Gray- Mexican Holdback

Nutritional value

Table 5.2, 5.3 and 5.4 are listed the nutritional characteristics of the leaves of *C. mexicana* collected in northeastern Mexico. It has high organic matter and medium value of crude protein. The neutral detergent fiber is low and its components (cellulose, hemicellulose and lignin). The hemicellulose has an arbitrary, unstructured form with very little strength. Hemicellulose may contain different amounts of arabinose, galactose, mannose, rhamnose, and xylose. Hemicellulose is a structural carbohydrate that can be digested to some degree by rumen microorganisms to give energy to the ruminant animal. Rumen bacteria digest structural carbohydrates more gradually than nonstructural carbohydrates (e.g. sugars and starches).

The leaves of *C. mexicana* do not contain condensed tannins. However, it has high molar proportions of acetic acid followed by propionic acid and butyric acid. The predicted dry matter digestibility, digestible energy, metabolizable energy and dry matter intake are high to meet adult ruminant metabolic needs, in all seasons of the year. The Na, P, Mg, Cu and Zn were marginal lower to satisfy the adult ruminant requirements; however, Ca, K, Mg and Fe resulted sufficient.

Table 5.2. Seasonal changes of the chemical composition (%, dry matter) and molar proportions of volatile fatty acids (mol/100 Mol, dry matter) in leaves of *Caesalpinia mexicana* collected in northeastern Mexico

Concept	Seasons				Annual
	Winter	Spring	Summer	Fall	mean
Organic matter	85	91	91	87	85
Ash	15	9	10	14	12
Crude protein	13	14	18	14	14
Degradable protein	8	9	12	10	9
Undegraded protein	5	5	6	4	5
Neutral detergent fiber	24	30	33	24	28
Insoluble neutral detergent fiber	7	8	9	7	8
Cellular content	76	70	67	76	72
Acid detergent fiber	18	21	25	18	20
Cellulose	12	14	15	10	13
Hemicellulose	6	9	8	6	7
Lignin	6	7	10	8	8
Condensed tannins	0	0	1	1	1
Acetic acid	74	69	73	73	72
Propionic acid	20	25	21	20	22
Butyric acid	6	6	6	7	6

Table 5.3. Seasonal variation of predicted dry mater digestibility (%), digestible energy (Mcal/kg), metabolizable energy (Mcal/kg) and dry matter intake (g/kg BW/day) of the leaves of *Caesalpinia mexicana* collected in northeastern Mexico

Concept	Seasons				Annual
	Winter	Spring	Summer	Fall	mean
Dry matter digestibility	71	69	66	71	69
Digestible energy	3.3	3.2	3.1	3.3	3.2
Metabolizable energy	2.7	2.6	2.5	2.7	2.7
Dry matter intake	84	84	84	84	84

Table 5.4. Seasonal content of macro (g/kg, dry matter) and microminerals (mg/kg, dry matter) in leaves of *Caesalpinia mexicana* collected in northeastern Mexico

Concept	Seasons				Annual mean
	Winter	Spring	Summer	Fall	
Macrominerals					
Ca	45	18	29	39	33
K	3	6	9	5	6
Mg	2	2	2	2	2
Na	<1	<1	<1	<1	<1
P	1	1	1	1	1
Microminerals					
Cu	4	4	4	4	4
Fe	61	71	41	59	58
Mn	50	26	83	65	56
Zn	29	26	48	26	32

CHAPTER 6

Nutrition of *Calliandra eriophylla* Benth.

Introduction

The plant species is known as the Fairy Duster that commonly propagates in thirsty regions. It may subsist with slight water, and with diverse quantities of contact to the sun. Nonetheless, the species flowers more when it is visible to the sun than when it has little visible to the sun. It can be governed by on soil that is dry and comprises gravel, and it is certainly alkaline. It is dispersed across a extensive range of dry and arid regions. It is very frequently situated in the Southwestern United States of America and Northern Mexico. The flowers are a pale pink color that are a fascination to ruminants that habit in the desert. This species serves a source of feed for the deer that browse in the dry areas or hillsides regions. The taxonomic characteristic of *Calliandra eriophylla* are listed in Table 6.1.

Calliandra eriophylla

Table 6.1. Taxonomic characteristics of *Calliandra eriophylla*

Rank	Scientific Name and Common Name
Kingdom	Plantae – Plants
Subkingdom	Tracheobionta – Vascular plants
Superdivision	Spermatophyta – Seed plants
Division	Magnoliophyta – Flowering plants
Class	Magnoliopsida – Dicotyledons
Subclass	Rosidae
Order	Fabales
Family	Fabaceae/Leguminosae – Pea family
Genus	Calliandra Benth. – stickpea
Species	*Calliandra eriophylla Benth. – fairyduster*

Nutritional value

In Tables 6.2, 6.3 and 6.4 are listed the nutritional characteristics of leaves of *Calliandra eriophylla*, consumed by range ruminates and collected at the north region of Durango State, Mexico. It has a medium value of crude protein, low neutral detergent fiber, hemicellulose, lignin and cellulose are almost similar, high values of condensed tannins and low ether extract, the acetic acid is much higher than propionic or butyric, isobutyric, isovaleric valeric acids. Except of Ca and Fe, other minerals insufficient to satisfy the adult small ruminant metabolic requirements. The *in vitro* gas production was low along with the bypass protein, microbial protein and metabolizable protein. The potential dry matter digestibility was also medium; however, the potential digestible energy, potential metabolizable energy and potential dry matter intake values were sufficient for the needs of range ruminant consuming leaves of *Calliandra eriophylla*.

Table 6.2. Chemical composition (%) and *in vitro* volatile fatty acids (mM) of leaves from *Calliandra eriophylla* collected in the state of Durango México

Concept	Dry matter basis
Ash	11
Crude protein	11
Neutral detergent fiber	43
Acid detergent fiber	29
Lignin	12
Hemicellulose	14
Cellulose	18
Condensed tannins	9
Ether extract	2
Acetic	13
Propionic	4
Butyric	2
Isobutyric	1
Isovaleric	1
Valeric	1
Total	21
Acetic:Propionic	3

Table 6.3. Mineral composition of macro (g/kg, dry matter) and microminerals (mg/kg, dry matter) of leaves from *Calliandra eriophylla* collected in the state of Durango México

Concept	Minerals
Macrominerals	
Ca	26
K	1
Mg	2
Na	0.3

P	1
Microminerals	
Cu	3
Fe	162
Mn	24
Zn	14

Table 6.4. *In vitro* gas production in 24 hours (GasP$_{24h}$/ mL 200 mg, dry matter), bypass protein (%, dry matter), microbial protein (%, dry matter) and metabolizable protein (%, dry matter) in leaves from *Calliandra eriophylla* collected in north of the state of Durango México

Concept	Dry matter
In vitro gas production	26
Bypass protein	3.6
Microbial protein	3.1
Metabolizable protein	6.7
Potential dry matter digestibility	61
Potential digestible energy, Mcal/kg	2.9
Potential metabolizable energy, Mcal/kg	2.4
Potential dry matter intake, g/kg/day	83

CHAPTER 7

Nutrition of *Desmanthus virgatus* (L.) Willd.

Introduction

The species is native to Mexico. It belongs to the family Leguminoseae (Fabaceae) it is known in Mexico as huizachillo. In the USA, ground tamarind and mimosa. It is a perennial legume that grows in summer on clay soils that receive 500 to 750 mm of water; it is resistant to neutral pH and grows in large areas of grasslands. It is productive and drought tolerant. It grows in tropics and subtropics, still reasonably cold tolerant and although it is defoliated by frost, this regrowth of crowns once there is enough moisture. Under heavy grazing conditions, all varieties of *Desmanthus* adopt a habit of rosette-crushed. It is mainly used as feed because of its high protein content and low toxicity to animals that consume it.

Desmanthus virgatus

65

Table 7.1. Taxonomic characteristics of *Desmanthus virgatus*

Kingdom	Plantae – Plants
Sub kingdom	Tracheobionta – Vascular plants
Super division	Spermatophyta – Plants with seeds
Division	Magnoliophyta – Plants with flowers
Class	Magnoliopsida – Dicotyledonous
Subclass	Rosidae –
Order	Fabales –
Family	Fabaceae – Leguminosae
Genus	*Desmanthus*
Species	*Desmanthus virgatus* (L.) Willd.

Nutritional value

The huizachillo is very appetizing and easily consumed by white-tailed deer. Studies have determined that the plant has a level of protein and digestibility equal to the Leucaena. It has a high protein and its use for low forage species and growth especially in heavy soils is recommended. The dry matter production is about 7.6 ton/ha/cut, in Hawaii can produce up to 23 ton/ha/cut and in Australia up to 70 ton/ha/ year. Hawaii plants are mechanically cut to a height of 5 to 7.5 cm. ground. Cutting intervals 91 days to obtain a dry matter yield of 23 ton/ha/year. The forage yield decreases with the progress of time; for example, after 4 years of implementation the plant mortality is advanced. *Leucaena* leucocephala exceeds *Desmanthus* regarding forage yield; however, it has an interesting feature; *Desmanthus* is not toxic to animals.

Table 7.2. Seasonal variation of the chemical composition (%, dry matter) and molar proportions of volatile fatty acids in leaves of *Desmanthus virgathus* collected in northeastern Mexico

Concept	Seasons				Annual
	Winter	Spring	Summer	Fall	mean
Organic matter	85	88	83	87	86
Ash	15	12	17	13	14
Crude protein	22	18	19	22	21
Degraded protein	12	10	12	13	12

Undegraded protein	10	8	7	9	9
Neutral detergent fiber	34	33	39	36	34
Insoluble neutral detergent fiber	10	8	7	9	9
Cellular content	66	67	61	64	66
Acid detergent fiber	28	26	31	29	28
Cellulose	9	6	8	7	9
Hemicellulose	8	7	8	7	8
Lignin	11	11	12	12	12
Condensed tannins	7	9	7	12	9
Acetic acid	67	68	59	72	66
Propionic acid	27	28	37	22	29
Butyric acid	7	5	4	5	5

Table 7.3. Seasonal variation of predicted dry mater digestibility (%), digestible energy (Mcal/kg), metabolizable energy (Mcal/kg) and dry matter intake (g/kg BW/day) of the leaves of *Desmanthus virgatus* collected in northeastern Mexico

Concept	Seasons				Annual
	Winter	Spring	Summer	Fall	mean
Dry matter digestibility	64	65	61	63	64
Digestible energy	3.0	3.1	2.9	3.0	3.0
Metabolizable energy	2.6	2.6	2.5	2.5	2.6
Dry matter intake	83	84	83	83	83

Table. 7.4. Seasonal content of macro (g/kg) and microminerals (mg/kg) of leaves of *Desmanthus virgathus* collected at northeastern Mexico

Concept	Seasons				Annual mean
	Winter	Spring	Summer	Fall	
Macrominerals					
Ca	14	9	16	12	13
K	9	8	6	12	9
Mg	9	9	8	5	7
Na	<1	<1	<1	<1	<1
P	1	1	1	1	1
Microminerals					
Cu	5	7	16	6	9
Fe	456	420	433	459	442
Mn	74	59	58	54	61
Zn	28	29	30	25	27

CHAPTER 8

Nutrition of *Ebenopsis ebano* (Berland.) Barneby & J.W.Grimes

Introduction

Ebanopsis ebano belongs to the Leguminoseae (Fabaceae) family that is native to of southern Texas, USA and eastern Mexico. It is known as Texas Ebony or Ébano in Spanish. It has been used since the arrival of the first settlers in northeastern Mexico; because of its hardness, durability and longevity represented good option in building houses, fences, furniture, tools, and crafts and as an ornamental element. The stalks are a source of firewood, coal and shelves. The leaves and young fruits are used as fodder. Humans consume the seeds of fruit when they are tender and raw, roasted or boiled in water; mature seeds are ground roasted and mixed with coffee, they are also widely used in jewelry. The taxonomic characteristics of *E. ebano* are listed in Table 8.1.

Ebanopsis ebano

Table 8.1. Taxonomic characteristics of *Ebenopsis ebano*

Kingdom	Plantae – Plants
Sub kingdom	Tracheobionta – Vascular plants
Super division	Spermatophyta – Plants with seeds
Division	Magnoliophyta – Plants with flowers
Class	Magnoliopsida – Dicotyledonous
Subclass	Rosidae –
Order	Fabales –
Family	Fabaceae – Leguminosae
Genus	*Ebenopsis*
Species	*ebano* (Berland.) Barneby & J.W.Grimes

Nutritional value

The nutritional profile of leaves from *Ebanopsis ebano* is shown in Tables 8.2, 8.4 and 8.4. The organic matter and ash contents were high in all season. Even though crude protein content was lower during spring, in all seasons resulted sufficient to meet the crude protein requirements of adult range ruminants. The neutral detergent fiber (NDF) and acid detergent fiber (ADF) content were lower than grasses; however, lignin content, in all seasons was high. The content of indigestible NDF (INDF) that requires extended *in situ* incubation periods represents a significant indicator of the nutritional quality of plant carbohydrates and may be a good predictor of *in vivo* digestion of forages. The amount of NDF may affect the digestibility of the forages in ruminants. The INDF is the main feature disturbing the dietary organic matter digestibility. The INDF is inaccessible to digestion of ruminal microbes, and INDF contents significantly augmented through maturation of plants that may have practical consequences for the period of cut.

The condensed tannins concentrations were low during all seasons. The molar proportions of the volatile fatty acids were higher during summer compared to other seasons. The acetic acid was twice as much as propionic acid and butyric acid was lowest. The predicted dry matter digestibility, digestible energy, metabolizable energy and dry matter intake of leaves from *E. ebano* were also higher during summer than other seasons of the year. The Mg, Na, P, Cu, Fe and Zn resulted marginal lower to fulfill the needs of adult range ruminants. However, Ca, K and Mn were in sufficient amounts.

Table 8.2. Seasonal tendency of the chemical composition (%, dry matter) and molar proportions of volatile fatty acids in leaves of Ebenopsis ebano collected in northeastern Mexico

Concept	Seasons				Annual
	Winter	Spring	Summer	Fall	mean
Organic matter	91	90	90	89	90
Ash	9	10	10	11	10
Crude protein	22	15	25	23	21
Neutral detergent fiber	49	39	56	55	50
Insoluble acid detergent fiber	12	9	14	12	11
Cellular content	51	61	44	45	50
Acid detergent fiber	32	23	38	32	31
Cellulose	12	12	14	11	12
Hemicellulose	17	17	18	23	19
Lignin	19	11	23	24	20
Condensed tannins	1	2	1	1	1
Acetic acid	67	50	70	53	61
Propionic acid	26	30	28	27	36
Butyric acid	7	6	2	5	5

Table 8.3. Seasonal variation of predicted dry mater digestibility (%), digestible energy (Mcal/kg), metabolizable energy (Mcal/kg) and dry matter intake (g/kg BW/day) of the leaves of Ebenopsis ebano collected in northeastern Mexico

Concept	Seasons				Annual
	Winter	Spring	Summer	Fall	mean
Dry matter digestibility	61	56	67	61	61
Digestible energy	2.9	2.7	3.1	2.9	2.9
Metabolizable energy	2.4	2.2	2.6	2.4	2.4
Dry matter intake	82	81	83	82	82

Table. 8.4. Seasonal variations of macro (g/kg, dry matter) and microminerals (mg/kg, dry matter) of the leaves of *Ebenopsis ebano* collected in northeastern Mexico

Concept	Seasons				Annual mean
	Winter	Spring	Summer	Fall	
Macrominerals					
Ca	18	22	22	25	22
K	13	11	15	13	13
Mg	3	4	3	3	3
Na	<1	<1	<1	<1	<1
P	1	1	2	1	1
Microminerals					
Cu	6	5	6	6	6
Fe	38	37	33	40	37
Mn	57	33	41	53	46
Zn	23	15	16	23	19

CHAPTER 9

Nutrition of *Eysenhardtia polystachya* (Ortega) Sarg.

Introduction

Eysenhardtia polystachya belongs to the family Fabaceae (Leguminoceae), it is commonly known in Mexico as vara dulce. It is a plant native to northern Mexico and is widely distributed on both sides and in the central part of the country at an altitude of 150 to 3.000 m. Its foliage is deciduous, flowers from May to October and fruits from November to December. Species widely used for firewood, because it has good energy characteristics. It provides fodder in abundance. It is highly desirable by cattle and goats that browse the twigs. Taxonomic characteristics of *Eysenhardtia polystachya* are listed in Table 9.1.

Eysenhartia polystachya

Table 9.1. Taxonomic characteristics of *Eysenhartia polystachya*

Kingdom	Plantae – Plants
Sub kingdom	Tracheobionta – Vascular plants
Super division	Spermatophyta – Plants with seeds
Division	Magnoliophyta – Plants with flowers
Class	Magnoliopsida – Dicotyledonous
Subclass	Rosidae –
Order	Fabales –
Family	Fabaceae – Leguminosae
Genus	*Eysenhartia*
Species	*polystachya* (Ortega) Sarg. Vara dulce

Nutritional value

In Tables 9.2, 9.3 and 9.4 is shown the nutritional value of *E. polystachya*. It can be pointed out that has high organic matter, crude protein content and degraded protein. However, it has low neutral detergent fiber and its components (cellulose and hemicellulose. The cellulose is a carbohydrate that is part of the primary cell wall of the plant. In addition, cellulose is the most abundant compound from organic origin on Earth. Ruminants may digest cellulose helped by symbiotic microorganisms that are hosted in their stomachs. Ruminants like goats, cows, sheep and white-tailed deer hold several symbiotic anaerobic bacteria in the rumen, and the bacteria produce enzymes called cellulases that aid the microorganisms to digest cellulose; the bacteria for propagation eventually utilize the digested products. The ruminant in its digestive system (stomach and small intestine) later digests the bacterial mass. The leaves of *Eysenhardtia polystachya* have no condensed tannins. The acetic acid is twice as much as propionic acid and butyric acid is lowest. The predicted dry matter digestibility, digestible energy, metabolizable energy and dry matter intake are high compared to commercial feeds such as alfalfa hay. The Na, P and Cu content resulted lower for adult ruminants consumed this plant; however, Ca, K, Mg, Fe, Mn and Zn were in sufficient amounts to adult ruminant needs.

Table 9.2. Seasonal variation of the chemical composition (%, dry matter) and molar proportions of volatile fatty acids (Mol/100 mol, dry matter) of leaves from *Eysenhardtia polystachya* collected in northeastern Mexico

Concept	Seasons				Annual
	Winter	Spring	Summer	Fall	mean
Organic matter	93	93	86	91	91
Ash	7	7	14	9	9
Crude protein	21	23	17	22	21
Degradable protein	15	18	10	15	14
Undegraded protein	6	5	7	7	7
Neutral detergent fiber	36	29	33	39	34
Insoluble neutral detergent fiber	9	6	6	8	7
Cellular content	64	71	67	61	66
Acid detergent fiber	23	16	16	21	19
Cellulose	12	11	8	11	11
Hemicellulose	13	12	17	18	15
Lignin	11	10	9	9	8
Condensed tannins	0	1	0	0	0
Acetic acid	67	67	74	68	69
Propionic acid	27	27	21	26	25
Butyric acid	6	6	5	6	6

Table 9.3. Seasonal variation of predicted dry mater digestibility (%), digestible energy (Mcal/kg), metabolizable energy (Mcal/kg) and dry matter intake (g/kg BW/day) of the leaves of *Eysenhartia polystachya* collected in northeastern Mexico

Concept	Seasons				Annual
	Winter	Spring	Summer	Fall	mean
Dry matter digestibility	68	74	73	70	71
Digestible energy	3.2	3.4	3.4	3.3	3.3
Metabolizable energy	2.6	2.8	2.8	2.7	2.7
Dry matter intake	83	84	84	83	83

Table 9.4. Season al content of macro (g/kg, dry matter) and microminerals (mg/kg, dry matter) in leaves of *Eysenhardtia polystachya* collected in northeastern Mexico

Concept	Seasons				Annual
	Winter	Spring	Summer	Fall	Media
Macrominerals					
Ca	22	23	34	22	25
K	6	10	5	10	8
Mg	3	3	2	3	3
Na	<1	<1	<1	<1	<1
P	1	1	1	1	1
Microminerals					
Cu	5	5	7	4	5
Fe	88	124	109	107	107
Mn	50	37	81	62	57
Zn	70	66	119	85	85

CHAPTER 10

Nutrition of *Havardia pallens* (Benth.) Britt. & rose

Introduction

Havardia pallens belongs to the <u>Leguminoseae</u> (<u>Fabaceae</u>) family. It belongs to the subfamily <u>Mimosoideae</u>. This plant is native to Mexico, where its extension covers the northeastern part of Mexico and part of southern Texas and Arizona. It has ornamental use as a small thorny tree very decorative delicate pinnate leaves composed. It spreads by seed, in clay soils that are well drained, and it is not drought tolerant as other plants. Requires little maintenance and the amount of sun can be full sun to partial shade. The importance of *H. pallens* is that it is one of the main bushes consumed by goats; in addition, provides a source a protective cover for small birds and a good place to nest.

Havardia pallens

In the United States of America, it is known by the name of "ape's earring". *Pallens* synonymy is *Calliandra* Benth., *Feuilleea brevifola* (Benth.) Kuntze., *Havardia brevifolia* (A. Gray) Small., *Havardia nelsonii* Britton & Rose., *Phitecellobium pallens* Britton & Rose., *Pithecellobium brevifolium* A. Gray., *Pithecellobum nelsonii* (Britton & Rose) Standley., *Pithecellobium brevifolium* A. Gray. and *Zygia brevifolia* (Benth.) Sudw (12). I is a shrub 1-3 m tall; occasionally trees 5 to 8 m. often not until well forked overhead. The foliage is pale green. The gray-brown bark is thorny; has bipinnately compound leaves, leaflets 4-5 mm long. The flowers occur in globular, white sets; fragrant after rain; the fruit is a seedpod directly, thin, flat wall. It reddish brown color. Its growth rate is slow (2). The flowering season is between March and April. In this region, white flowers release a wonderful sweet fragrance. The fruit is oblong, 5-15 cm long and 1 to 1.8 cm wide, it is flat, straight, dehiscent, and green to dark brown; membranous, thick and glabrous valves. The taxonomic characteristics are listed in Table 10.1

Table 10.1. Taxonomic characteristics of *Havardia pallens*

Kingdom	Plantae – Plants
Sub kingdom	Tracheobionta – Vascular plants
Super division	Spermatophyta – Plants with seeds
Division	Magnoliophyta – Plants with flowers
Class	Magnoliopsida – Dicotyledonous
Subclass	Rosidae –
Order	Fabales –
Family	Fabaceae – Leguminosae
Genus	*Havardia*
Species	*pallens* (Berland.) (Benth.) Britton & Rose – haujillo

Nutritional value

In Table 10.2, 10.3 and 10.4 are listed the nutritional characteristics of the leaves of *Havardia pallens*. This plant has high organic matter. The crude protein content is also high for the adult ruminant needs. It has low neutral detergent fiber and high cellular content. The cell content (cytoplasm) is a semitransparent liquid in which the other organelles are adjourned. In it takes the procedures such as breaking down feed compounds into smaller particles. The cytoplasm involves mainly of water, dissolved ions, minor particles, and great water-soluble particles such as proteins. In addition, it has a high content of K ions and a low content of Na ions that aid in regulatory the quantity of water in inside the cell. The leaves of *Havardia pallens* have low condensed tannins concentration. The acetic acid is almost as double as much of propionic acid and butyric acid is lowest. The predicted dry matter digestibility, digestible energy and metabolizable energy are sufficient for the adult ruminant requirements. It has also high levels of dry matter intake. The concentrations of P, Na, Cu and Zn were marginal lower for adult range ruminants consuming this plant. Ca, Mg, K, Fe and Mn were sufficient.

Table. 10.2. Seasonal variations of the chemical composition (%, dry matter) and molar proportions of volatile fatty acids (Mol/100 Mol, dry matter) in leaves of *Havardia pallens* collected in northeastern Mexico

Concept	Seasons				Annual
	Winter	Spring	Summer	Fall	mean
Organic matter	91	90	93	90	91
Ash	9	10	7	10	9
Crude protein	22	18	25	19	21
Degradable protein	12	13	19	11	13
Undegraded protein	10	5	6	8	8
Neutral detergent fiber	36	29	39	40	36
Insoluble neutral detergent fiber	10	8	10	10	10
Cellular content	64	71	61	60	64
Acid detergent fiber	27	20	28	28	26
Cellulose	17	15	18	18	17
Hemicellulose	10	9	12	12	11
Lignin	8	4	9	9	7
Condensed tannins	1	0	1	0	1
Acetic acid	70	67	71	67	69
Propionic acid	20	22	19	24	22
Butyric acid	10	11	10	7	9

Table 10.3. Seasonal variation of predicted dry mater digestibility (%), digestible energy (Mcal/kg), metabolizable energy (Mcal/kg) and dry matter intake (g/kg BW/day) of the leaves of *Havardia pallens* collected in northeastern Mexico

Concept	Seasons				Annual
	Winter	Spring	Summer	Fall	mean
Dry matter digestibility	65	70	65	64	66
Digestible energy	3.0	3.3	3.0	3.0	3.1
Metabolizable energy	2.5	2.7	2.5	2.5	2.5
Dry matter intake	83	84	83	83	83

Table 10.4. Seasonal content of macro (g/kg, dry matter) and microminerals (mg/kg, dry matter) in leaves of *Havardia pallens* collected in northeastern Mexico

Concept	Seasons				Annual
	Winter	Spring	Summer	Fall	mean
Macrominerals					
Ca	18	25	17	25	21
K	9	9	8	5	8
Mg	4	3	3	4	4
Na	<1	<1	<1	<1	<1
P	1	1	1	1	1
Microminerals					
Cu	6	5	5	6	6
Fe	107	86	79	132	100
Mn	64	33	44	60	50
Zn	17	15	13	15	15

CHAPTER 11

Nutrition of *Leucaena leucocephala* (L.) Benth.

Introduction

The *L. leucocephala* belongs to the Leguminoseae (Fabaceae) family. In Mexico, it is known as Guaje and the United States is called Leucaena. Synonymy is *Leucaena glauca* (L.) Benth, *Acacia glauca* L., *Acacia glauca* Willd, *Acacia leucocephala* (Lam) Link, *Leucaena glabrata* Rose, *Leucaena latisiliqua* (L.) Gillis & Steam, and *Mimosa leucocephala* Lam. The plant is perhaps the most widely studied in the and certainly in Mexico. Studies on this plant include assessments establishment, agronomic performance, and response of cattle under grazing conditions, chemical composition and digestibility. These studies showed that the Leucaena is a good nutritional value as a forage to increase livestock production in areas subtropical in regions where this plant grows. *Leucaena leucocephala* is a multipurpose tree of great interest in agroforestry. It is often mixed with agricultural crops. As a food source in animal production farms, among its benefits are high crude protein, vitamins, minerals, carotene and fiber that determine its high nutritional value. However, it tolerates moderate drought periods, and have ability to adapt to moderately acidic soils (pH 5.0 to 6.4). Thus, because of its high productive potential it may be considered a good alternative to improve livestock farms.

Nutritional value

Its foliage (leaves, stems, seeds and fruits) is consumed by wild and domestic ruminants, the leaves are an excellent fodder (30% dry matter,

20-27% protein, and rich in Ca, K and vitamins). Digestibility have a percentage of 60 to 70%. The leaves and seeds contain a toxic amino acid (mimosine) which may cause damage to nonruminant animals and poultry, causing weakness, weight loss, abortion and hair loss in horses, mules and donkeys. Ruminants counteract the toxic effect by ruminal bacteria.

Leucaena leucocephala

Table 11.1. Taxonomic characteristics of *Leucaena leucocephala*

Kingdom	Plantae – Plants
Sub kingdom	Tracheobionta – Vascular plants
Super division	Spermatophyta – Plants with seeds
Division	Magnoliophyta – Plants with flowers
Class	Magnoliopsida – Dicotyledonous
Subclass	Rosidae –
Order	Fabales –
Family	Fabaceae – Leguminosae
Genus	*Leucaena*
Species	*leucocephala* (L.) Benth. - Leucaena

In Tables 11.2, 11.3, 11.4 and 11.5 are listed the nutrition data from the leaves of *L. leucocephala* collected in northeastern Mexico at different sites and dates.

Table 11.2. Seasonal variation of the chemical composition (% dry matter) and the proportions of volatile fatty acids (Mol/100 Mol) in leaves of *Leucaena leucocephala* collected in northeastern Mexico

| Concept | Seasons | | | | Annual |
	Winter	Spring	Summer	Fall	mean
Organic matter	91	90	91	91	91
Ash	9	10	9	9	9
Crude protein	20	25	26	23	24
Degradable protein	12	15	17	15	15
Undegraded protein	8	10	9	8	9
Neutral detergent fiber	47	43	44	44	45
Insoluble neutral detergent fiber	6	6	6	7	6
Cellular content	53	57	56	56	55
Acid detergent fiber	19	15	15	17	15
Cellulose	6	7	6	7	7
Hemicellulose	28	28	29	27	30
Lignin	7	10	9	9	9
Condensed tannins	6	7	6	11	8
Acetic acid	67	66	67	74	69
Propionic acid	27	27	27	22	26
Butyric acid	6	5	6	5	6

Table 11.3. Chemical composition (%, dry matter basis) of *Medicago sativa* hay, and leaves of *Leucaena leucocephala* collected in Cienega de Flores county of the state of Nuevo Leon, Mexico

Concept	M. sativa hay	L. leucocephala leaves
Crude protein	17.5	25.0
Ashes	10.8	10.6
Neutral detergent fiber	35.8	23.8
Acid detergent fiber	31.1	13.2
Cellulose	23.4	10.7
Hemicellulose	9.5	10.5
Acid detergent lignin	5.7	2.4
Condensed tannins,	0.1	4.5

Dry matter digestibility	61	77
Digestible energy (Mcal/kg)	2.9	3.6
Metabolizable energy (Mcal/kg)	2.4	3.0
Dry matter intake (g/kg BW/day)	83	84

Table 11.4. Seasonal variation of predicted dry mater digestibility (%), digestible energy (Mcal/kg), metabolizable energy (Mcal/kg) and dry matter intake (g/kg BW/day) of the leaves of *Leucaena leucocephala* collected in northeastern Mexico

Concept	Seasons				Annual
	Winter	Spring	Summer	Fall	mean
Dry matter digestibility	71	75	75	73	75
Digestible energy	3.3	3.5	3.5	3.4	3.5
Metabolizable energy	2.8	3.0	3.0	2.9	3.0
Dry matter intake	82	83	83	83	82

Table 11.5. Seasonal content of macro (g/kg, dry matter) and microminerals (mg/kg, dry matter) in leaves *Leucaena leucocephala* collected in northeastern Mexico

Concept[1]	Seasons				Annual
	Winter	Spring	Summer	Fall	mean
Macrominerals					
Ca	12	10	14	13	12
K	16	17	14	18	17
Mg	6	7	5	6	7
Na	1	1	1	1	1
P	1	1	1	1	1
Microminerals					
Cu	7	12	7	5	8
Fe	103	126	267	99	149
Mn	37	26	27	31	30
Zn	24	21	20	19	21

CHAPTER 12

Nutrition of *Mimosa aculeaticarpa* Ortega

Introduction

Livestock seldom browse the species; nonetheless, range goats and white-tailed deer may be used casually if other fodders are infrequent. Livestock usually consume the pods. The seeds are eaten by Scaled and Gambel's quail. It is of lower preference as a browse species for pronghorn and deer. It has has dense spikes and a twisted growth form which may be the reason for its low acceptability to livestock. However, the pods are very acceptable to cattle. Thus, the acceptability of the plant has been rated as poor for cattle and sheep and good for pronghorn. The taxonomic characteristics of *Mimosa aculeaticarpa* are listed in Table 12.1

Mimosa aculeaticarpa

Table 12.1. Taxonomic characteristics of *Mimosa aculeaticarpa*

Rank	Scientific Name and Common Name
Kingdom	Plantae – Plants
Subkingdom	Tracheobionta – Vascular plants
Superdivision	Spermatophyta – Seed plants
Division	Magnoliophyta – Flowering plants
Class	Magnoliopsida – Dicotyledons
Subclass	Rosidae
Order	Fabales
Family	Fabaceae/Leguminosae – Pea family
Genus	*Mimosa L.* – sensitive plant
Species	*Mimosa aculeaticarpa* Ortega – catclaw mimosa

Nutritional value

In Tables 12.1, 12.2 and 12.3 are listed the nutritional characteristics of leaves of *Mimosa aculeaticarpa*, consumed by range ruminates and collected at the north region of Durango State, Mexico. It has high crude protein content, low neutral detergent fiber. The cellulose content is higher than hemicellulose or lignin. It has high values of condensed tannins and low ether extract, the acetic acid is low; however, is much higher than propionic or butyric, isobutyric, isovaleric valeric acids. Except of Mg, Na, Cu, Fe and Mn, other minerals were sufficient to fulfill the adult small ruminant metabolic requirements. The *in vitro* gas production was low along with the bypass protein, microbial protein and metabolizable protein. The potential dry matter digestibility was also medium; however, the potential digestible energy, potential metabolizable energy and potential dry matter intake values were sufficient for the needs of range ruminant consuming leaves of *Mimosa aculeaticarpa*.

Table 12.2. Chemical composition (%) and *in vitro* volatile fatty acids (mM) of leaves from *Mimosa aculeaticarpa* collected in the state of Durango México

Concept	Dry matter basis
Ash	6
Crude protein	16
Neutral detergent fiber	47
Acid detergent fiber	33
Lignin	14
Hemicellulose	14
Cellulose	19
Condensed tannins	8
Ether extract	1
Acetic	10
Propionic	2
Butyric	1
Isobutyric	0.4
Isovaleric	0.2
Valeric	0.4
Total	14
Acetic:Propionic	5

Table 12.3. Mineral composition of macro (g/kg, dry matter) and microminerals (mg/kg, dry matter) of leaves from *Mimosa aculeaticarpa* collected in the state of Durango México

Concept	Minerals
Macrominerals	
Ca	15
K	4
Mg	2
Na	1
P	2

Microminerals	
Cu	3
Fe	44
Mn	21
Zn	33

Table 12.4. *In vitro* gas production in 24 hours (GasP$_{24h}$/ mL 200 mg, dry matter), bypass protein (%, dry matter), microbial protein (%, dry matter) and metabolizable protein (%, dry matter) in leaves from *Mimosa aculeaticarpa* collected in north of the state of Durango México

Concept	Dry matter
In vitro gas production	20
Bypass protein	3
Microbial protein	6
Metabolizable protein	9
Potential dry matter digestibility	59
Potential digestible energy, Mcal/kg	2.8
Potential metabolizable energy, Mcal/kg	2.3
Potential dry matter intake, g/kg/day	82

CHAPTER 13

Nutrition of *Parkinsonia aculeata* L.

Introduction

Parkinsonia aculeata is commonly known as Paloverde, It belongs to Leguminoseae (Fabaceae) family. It is native to Texas, and to west Arizona, USA. It is a very fast developing small shrub and tree. It grows in soils of low organic matter content. The plant grows under dry conditions, high saline conditions or high heat. This species necessitates plenty sun. The foliage and pods are frequently utilized as emergency fodder for livestock, as well as for wildlife. Bees from the flowers can yield aromatic honey. The taxonomic characterizes of *Parkinsonia aculeata* are listed in Table 13.1.

Parkinsonia aculeata

Table 13.1. Taxonomic characteristics of *Pakinsonia aculeata*

Kingdom	Plantae – Plants
Sub kingdom	Tracheobionta – Vascular plants
Super division	Spermatophyta – Plants with seeds
Division	Magnoliophyta – Plants with flowers
Class	Magnoliopsida – Dicotyledonous
Subclass	Rosidae –
Order	Fabales –
Family	Fabaceae – Leguminosae
Genus	*Parkinsonia*
Species	*Parkinsonia aculeata* L. palo verde.

Nutritional value

The nutritional profile of leaves from *Parkinsonia aculeata* are listed in Tables 13.2, 13.3 and 13.4. The leaves contain high organic matter content, crude protein and degraded protein. It appears that the most relevant measurement features responsible for the nutritional value of dietary protein, the nitrogen supply, peptides and short-chain acids to microbes in the rumen are the passage rate of undegradable proteins to the lower track. Some techniques have been utilized for the partition of crude protein into rumen degradable protein (RDP) and rumen undegradable protein (RUP). These techniques are *in situ* and *in vivo* estimations, and a diversity of *in vitro* techniques.

The aptitude of ruminants to produce and use high excellence microbial protein from inferior quality foods is the newest revolutionary adaptation of mammalian to the diets. To enhance progress and condition of the ruminant under confinement, feeds need to be elaborated to give well-adjusted levels of rumen degradable energy and nitrogen (RDN) to optimize microbial growth while providing sufficient rumen undegradable protein (RUP) for the ruminant.

The leaves of *P. aculeata* are also characterized by low neutral detergent fiber; cellulose is higher than hemicellulose or lignin. The plant practically do not has condensed tannins. The acetic acid is double as much as propionic acid and butyric acid is lowest. The predicted dry matter digestibility, digestible energy and metabolizable energy are high enough to meet adult ruminant requirements. The dry matter intake is also sufficient for adult range ruminants.

Table 13.2. Seasonal variation of the chemical composition (%, dry matter) and molar proportions of volatile fatty acids in leaves of *Parkinsonia aculeata* collected in northeastern Mexico

Concept	Seasons				Annual
	Winter	Spring	Summer	Fall	mean
Organic matter	89	93	94	92	92
Ash	11	7	6	8	8
Crude protein	17	19	22	17	19
Degraded protein	12	14	15	10	13
Undegraded protein	5	5	7	7	6
Neutral detergent fiber	46	49	49	53	49
Insoluble neutral detergent fiber	12	12	13	13	12
Cellular content	54	51	51	47	51
Acid detergent fiber	32	33	37	36	34
Cellulose	21	27	22	23	23
Hemicellulose	14	16	13	18	15
Lignin	11	12	15	12	13
Condensed tannins	0	0	0	0	0
Acetic acid	69	62	67	70	67
Propionic acid	24	30	31	26	27
Butyric acid	7	8	2	4	6

Table 13.3. Seasonal variation of predicted dry mater digestibility (%), digestible energy (Mcal/kg), metabolizable energy (Mcal/kg) and dry matter intake (g/kg BW/day) of the leaves of *Parkinsonia aculeata* collected in northeastern Mexico

Concept	Seasons				Annual
	Winter	Spring	Summer	Fall	mean
Dry matter digestibility	60	59	57	57	59
Digestible energy	2.8	2.8	2.7	2.7	2.8

Metabolizable energy	2.3	2.3	2.2	2.2	2.3
Dry matter intake	82	82	82	82	82

Table 13.4. Seasonal content of macro (g/kg, dry matter) and microminerals (mg/kg, dry matter) in leaves of *Parkinsonia aculeata* collected in northeastern Mexico

Concept	Seasons				Annual
	Winter	Spring	Summer	Fall	mean
Macrominerals					
Ca	28	13	13	22	19
K	13	11	16	8	12
Mg	2	2	2	2	2
Na	<1	<1	<1	<1	<1
P	0.9	1.5	1.7	1.3	1.4
Microminerals					
Cu	6	8	8	10	8
Fe	120	77	50	160	102
Mn	40	20	62	48	42
Zn	40	29	26	36	32

CHAPTER 14

Nutrition of *Parkinsonia texana* (A. Gray) S. Watson

Introduction

Parkinsonia texana is a small tree native to south Texas and northeastern Mexico. The plant is encountered on alkaline sandy loam or clays soils and is very tolerant to scarcity of water. The plant is a deciduous shrub, this is the shrub allows drop their leaves in times of drought to conserve water, and still be able to photosynthesize. *P. texana* grows at altitudes ranging from sea level up to 1200 meters. In greenhouse trials, the roots of grew an average of 0.9 centimeters per day. The plant is mainly used as fodder. Its nutritional value is very high and it is consumed by ruminants (white-tailed deer, goats, sheep, and cattle), each in different proportions, according to his availability. Small mammals eat the pods and seeds as they fall. The foliage ideal for nesting, which may offer *P. texana,* attracts dense and varied bird populations. The taxonomy characteristics are shown in Table 14.1.

Parkinsonia texana

Table 14.1. Taxonomy of *Parkinsonia texana*

Rank	Scientific Name and Common Name
Kingdom	Plantae- plants
Subkingdom	Tracheobionta – Vascular plants
Super division	Spermatophyta – Seed plants
Division	Magnoliophyta – Flowering plants
Class	Magnoliopsida – Dicotyledons
Subclass	Rosidae
Order	Fabales
Family	Fabaceae/Leguminosae – Pea family
Genus	*Parkinsonia* L. – paloverde
Species	*Parkinsonia texana* (A. Gray) S. Watson – Texas paloverde

Nutritional value

The ash content in leaves of *Parkinsonia texana* was very similar among seasons (Table 14.2). During summer and fall, the crude protein (CP) content was higher than other seasons. Higher CP seasonal ranges were also reported in other shrub growing in northeastern and were comparable to *Medicago sativa* hay. Because the high CP content, leaves from *P. texana* can be considered as good protein supplements for browsing range small ruminants. Lower percentages of neutral detergent fiber (NDF) content are mainly during winter and spring seasons, but in summer and fall were high (Table 14.2).

It has been reported that with exception of *Vachellia rigidula*, all evaluated shrubs growing in northeastern Mexico had annual mean NDF content lower than *M. sativa* hay. Cellulose values were low during winter and summer, but in spring and fall, the leaves of *P. texana* were high. The hemicellulose content in leaves of all plants was lower during summer than other seasons. The hemicellulose, in all seasons was higher than cellulose (Table 14.2). During spring lignin content was relatively low than in other seasons. The negative relationship between lignin concentration and digestibility has been already assumed. Digestion is limited when the polysaccharides are protected by lignin. Lignin has been negatively associated with the rate of NDF digestibility; thus, as lignin concentrations increase, an overall decrease in forage digestibility was detected. Moreover, the effect of lignin on fiber digestion is greater in grasses than in legumes.

Condensed tannins were higher in summer than in other seasons. Tannin content, potentially alter the use and value of tree foliages and may at times be responsible for the poor utilization of such forages by ruminant livestock. On the other hand, lack of tannins is believed to leave the protein so unprotected as to be completely degraded in the rumen. Tannins from *P. texana* appear to overprotect protein when could be fed to range small ruminants with consequent increased fecal loss of protein. The acetic acid content was very similar among seasons and was as double as much of propionic acid or butyric acid (Table 5.2). It has been reported lower volatile fatty acid values in oak leaves (22, 10 and 3 mmol/L, respectively).

Table 14.2. Seasonal variation of the chemical composition (%, dry matter) and volatile fatty acids (% molar, dry matter) content of leaves from *Parkinsonia texana* collected in northeastern Mexico

Concept[1]	Seasons				Annual
	Winter	Spring	Summer	Fall	mean
Organic matter	89	89	91	91	90
Ash	11	11	9	9	10

Crude protein	26	26	23	24	22
Degradable protein	14	14	16	17	12
Undegraded protein	10	9	10	9	10
Neutral detergent fiber	26	29	25	26	27
Insoluble neutral detergent fiber	6	7	5	5	6
Cellular content	74	71	75	74	73
Acid detergent fiber	14	15	15	14	15
Cellulose	7	9	5	6	7
Hemicellulose	10	12	12	14	12
Lignin	7	4	10	7	7
Condensed tannins	10	10	8	8	9
Acetic acid	71	74	72	73	73
Propionic acid	24	20	21	21	21
Butyric acid	5	6	7	6	6

Tree forages with a low NDF content (20-35%) are usually of high digestibility and species with high lignin contents are often of low digestibility. This effect is corroborated in data listed in Tables 14.1 and 14.2. When neutral detergent fiber was lower (summer and fall), the DMD was higher than in other seasons. The same effect was observed for the condensed tannins on DMD; season with lower condensed tannins content were higher in DMD, digestible energy, metabolizable energy and dry matter intake of leaves from Parkinsonia texana. In other studies, it has been published that the lignin content of tree foliage was negatively correlated ($r = -0.92$) with feed digestibility.

Table 14.2. Seasonal variation of digestion, energy content and intake of leaves from *Parkinsonia texana*

Concept	Seasons				Annual
	Winter	Spring	Summer	Fall	mean
Dry matter digestibility, %	78.2	75.5	81.2	80.4	78.8
Digestible energy, Mcal/kg	3.62	3.50	3.75	3.71	3.64
Metabolizable energy, Mcal/kg	2.97	2.87	3.08	3.05	2.99
Dry matter intake, g/kg BW/day	84.0	83.9	84.3	84.2	84.1

Dry matter digestibility was calculated as 83.58 - 0.824*acid detergent fiber + 2.626*nitrogen (Odd *et al.*, 1983).

Digestible energy calculated as 0.27+0.0428*DMD (Fonnesbeck *et al.*, 1984).

Metabolizable energy calculated as 0.821*DE (Khalil *et al.*, 1986).

IDNDF = indigestible neutral detergent fiber calculated as 86.98+1.542*NDF+31.63*Acid detergent fiber (Jancik *et al.*, 2008).

Dry matter intake was calculated as 86.5-0.09*neutral detergent fiber.

The concentration of individual minerals in forages varies greatly depending on soil, plant, and management factors. The stage of forage maturity affects the contents of minerals in forages. Nevertheless, limited quantitative information is available on seasonal dynamics of macro and trace elements in range plants consumed by range ruminants in semiarid regions. It seems that nutritional quality of range plants follows plant growth patterns. A peak generally occurs during spring when growth is most active and levels decline steadily reaching the lowest levels during winter. However, during years with abundant moisture coupled with mild winter temperatures, forage quality follows a bimodal pattern with peaks in quality occurring in spring and winter. All minerals listed in Table 14.3 varied seasonal. Seasonal variations in plant minerals might have been related to seasonal water deficits, excessive irradiance levels during summer and extreme low temperatures in winter that could have affected leaf development and senescence. In spite of these differences, all plant species had suitable levels of Ca, Mg, K, Fe and Mn to satisfy grazing ruminant requirements. However, Na, P, Cu and Mn, showed marginal inadequate concentrations in prolonged periods throughout the year and it might have a negative impact on range small ruminant productivity.

Table 14.3. Seasonal dynamics of macro (g/kg) and microminerals (mg/kg) contained in leaves of *Parkinsonia texana* collected in northeastern Mexico

Concept[1]	Seasons				Annual
	Winter	Spring	Summer	Fall	mean
Macrominerals					
Ca	11	11	14	11	12
K	13	17	18	16	16
Mg	8	10	14	11	12
Na	0.4	0.3	0.5	0.5	0.4

P	1	1	1	1	1
Microminerals					
Cu	5	5	4	5	5
Fe	57	83	171	102	103
Mn	20	24	29	25	24
Zn	58	50	61	50	55

In a study that consisted of two experiments, was evaluated the Chemical composition and the in vitro true organic matter digestibility (IVTOMD) of *Parkinsonia texana,* a leguminous species that grow in the Tamaulipan Thornscrub Vegetation, northeastern Nuevo Leon, Mexico. In the first experiment was tested the effects of season on chemical composition as nutritional variable to small ruminants. It seems that chemical composition was affected by season (Table 14.4).

Table 14.4. Seasonal variation of the chemical composition, *in vitro* true organic matter digestibility (IVTOMD, %) and gross energy losses (GE, % as methane production) of leaves of *Parkinsonia texana* collected in northeastern Nuevo Leon, Mexico.

Concept	NDF	CT	CP	EE	IVTOMD	GE, CH4
Spring	36.8	12.4	19.2	1.7	76.7	4.1
Summer	28.8	12.3	20.1	1.5	84.2	2.3
Autumn	32.8	6.6	22.2	1.5	84.2	2.3
Winter	32.6	7.4	21.6	1.6	82.5	2.7
Mean	32.8	9.7	20.8	1.6	81.9	2.9

NDF = neutral detergent fiber; CT = condensed tannins; CP = crude protein; EE = ether extract

Moreover, the IVTOMD was negatively correlated to NDF (r = - 0.86; P<0.001), condensed tannins content (r = - 0.56; P<0.01) and ether extract (r = - 0.95; P<0.001), and positively with crude protein (r = 0.71: P<0.001) and Gross energy (% of methane losses) (r = 0.93; P<0.001). In the second experiment it was tested the effect of the addition of polyethylene glycol (PEG) on in vitro fermentation parameters. Results listed in Table 6.5 showed that the addition of PEG increased fermentation parameters and metabolizable energy content (Table 14.5).

Table 14.5. Effect of polyethylene glycol on gas production (Gas24 h, ml/g dry matter), microbial protein (MP, µmol), partitioning factor (PF) and metabolizable energy (ME, Mcal/kg dry matter) of leaves of *Parkinsonia texana* in northeastern Nuevo Leon, Mexico

Concept	Gas24 h	Rate, h	Lag time, h	MP	PF	ME
Without PEG						
Spring	171	0.07	0.52	4.2	4.1	1.70
Summer	186	0.06	0.49	6.4	4.1	2.20
Autumn	188	0.07	1.45	9.6	4.3	2.40
Winter	187	0.08	1.01	4.8	4.4	2.00
Mean	183	0.07	0.87	6.4	4.2	2.08
With PEG						
Spring	193	0.08	1.13	5.7	4.8	1.90
Summer	208	0.08	1.28	7.3	4.8	2.20
Autumn	202	0.08	1.88	11.5	4.7	2.40
Winter	195	0.09	1.03	5.3	4.7	2.20
Mean	200	0.08	1.33	7.3	4.8	2.18

CHAPTER 15

Nutrition of *Prosopis glandulosa* Torr.

Introduction

Prosopis glandulosa belongs to the Fabaceae (Leguminosae) family is commonly known in Mexico as Mezquite. In the USA, it is known as honey mesquite, Western Money, mesquite and Texas mesquite. Mesquite plays a very important role in the ecosystem, helping in changing the environment extremes of deserts and arid areas, allowing other species to thrive otherwise would not develop in this type of conditions. It also provides food and shelter for many types of wildlife such as birds and mammals, thus, it achieves a valuable role in the conservation of wildlife, which represents a factor of support for the maintenance of complex ecosystems of arid and semiarid areas.

Prosopis glandulosa

Table 15.1. Taxonomic characteristics of *Prosopis glandulosa*

Kingdom	Plantae – Plants
Sub kingdom	Tracheobionta – Vascular plants
Super division	Spermatophyta – Plants with seeds
Division	Magnoliophyta – Plants with flowers
Class	Magnoliopsida – Dicotyledonous
Subclass	Rosidae –
Order	Fabales –
Family	Fabaceae – Leguminosae
Genus	*Prosopis*
Species	*Prosopis glandulosa* Torr. - Mezquite

Nutritional value

Livestock likes to feed the fallen leaves and fruits of Mezquite. It is a highly browsed by domestic (cattle, goats, sheep and small scale in horses, donkeys and mules) and wildlife (white-tailed deer). The mesocarp of mesquite seeds contains 13 to 36% of sugars mainly sucrose, the endosperm of the seed has 55 to 69 % protein, 8% fat with various minerals, all this gives it a higher nutritional value comparable o beans, peas, soybean, barley or corn. Chemical analysis green leaves of mesquite exposed similar nutritive value (fiber, crude protein, and gross energy) as mature alfalfa hay. However, there were identified, in a two studies, allelochemicals compounds that cause flavor aversions and other negative digestive concerns to animals that consume the fresh leaves of Mezquite.

Table. 15.2. Seasonal variation of the chemical composition (%, dry matter) and molar proportions of volatile fatty acids of leaves *Prosopis glandulosa* collected in northeastern Mexico

Concept	Seasons				Annual
	Winter	Spring	Summer	Fall	mean
Organic matter	92	93	88	87	90
Ash	8	7	12	13	10
Crude protein	19	19	22	18	20

Degraded protein	12	13	13	12	13
Undegraded protein	7	6	9	6	7
Neutral detergent fiber	42	47	41	38	42
Insoluble neutral detergent fiber					
Cellular content	58	53	59	62	58
Acid detergent fiber	31	35	31	27	31
Cellulose	16	19	17	13	17
Hemicellulose	11	12	10	11	11
Lignin	14	14	19	12	15
Condensed tannins	1	0	1	0	1
Acetic acid	66	64	65	75	65
Propionic acid	31	34	33	33	33
Butyric acid	3	2	2	2	2

Table 15.3. Seasonal variation of predicted dry mater digestibility (%), digestible energy (Mcal/kg), metabolizable energy (Mcal/kg) and dry matter intake (g/kg BW/day) of the leaves of *Prosopis glandulosa* collected in northeastern Mexico

Concept	Seasons				Annual
	Winter	Spring	Summer	Fall	mean
Dry matter digestibility	61	58	62	64	61
Digestible energy	2.9	2.7	2.9	3.0	2.9
Metabolizable energy	2.4	2.2	2.4	2.5	2.4
Dry matter intake	83	82	83	83	83

Table 15.3. Seasonal content of macro (g/kg, dry matter) and microminerals (mg/kg, dry matter) in leaves of *Prosopis glandulosa* collected in northeastern Mexico

Concept	Seasons				Annual
	Winter	Spring	Summer	Fall	mean
Macrominerals					
Ca	5	5	5	7	6
K	16	15	21	11	16
Mg	3	2	3	6	416

Na	<1	<1	<1	<1	<1
P	1	2	1	1	1
Microminerals					
Cu	4	5	3	4	4
Fe	92	173	108	93	166
Mn	39	24	36	42	47
Zn	27	58	47	62	49

CHAPTER 16

Nutrition of *Senegalia berlandieri* (Benth.) Britton & Rose

Introduction

Senegalia berlandieri (*Acacia berlandieri*) commonly known as Guajillo, is a legume small tree or shrub growing in northern Mexico and southern Texas. It belongs to the Leguminosae or Fabaceae family. It has seeds born in pods, compound leaves with numerous leaflets, and the roots are associated to bacteria that symbiotically fix nitrogen. This plant produce foliage and that are usually abundant in nitrogen compounds with a good indispensable amino acid composition. In addition. The plant provides food and good habitat protection for wildlife, and is browsed by both wildlife (white-tailed deer) and domestic (goats, sheep and cattle) ruminants, especially in periods of prolonged dry climate.

Senegalia berlandieri

However, guajillo contains antinutritional compounds such as phenolic amines, alkaloids, and condensed tannins that may induce toxicosis and diminish fertility, consumption, and nutrient digestion in ruminants. In a study that fed goats with mixed diets reported that nutrient digestibility reduced with augmented the amount of guajillo in diets. Moreover, guajillo did not meet digestible energy needs for maintenance. This fact comes to the question if the nutritional value of guajillo will be a good forage for white-tailed deer. Many plant secondary substances may have the possibility to cause negative effects on the performance of livestock. Plant antinutritive characteristics may be separated into a heat-labile group, including proteinase inhibitors, pectins and cyanogenic compounds that are delicate to regular treating temperatures and a heat constant group comprising, among several others, condensed tannins, antigenic proteins, saponins, mimosine and the nonprotein amino acids. These substances are present in the foliage and/or seeds of practically all plants that are used for food. During ruminal, digestion may diminish the effect of antinutritive compounds in some tree and shrub fodders for sheep, cattle, goats and white-tailed deer. Plants have evolved alone with killer of bacterial population, fungi, insects and browsing ruminants and have established protection mechanisms that assistance their existence. Many legume trees and shrubs frequently have spines, fiber foliage and growing conducts that defend the crown of the plant against defoliation. Mycotoxins synthetized by saprophytic and endophytic fungi are known as a potential cause of toxins in fodders.

Table 16.1. Taxonomic characteristics of *Senegalia berlandieri* (Benth.) Britton & Rose

Rank	Scientific Name and Common Name
Kingdom	Plantae- plants
Sub kingdom	Tracheobionta – Vascular plants
Super division	Spermatophyta – Seed plants
Division	Magnoliophyta – Flowering plants
Class	Magnoliopsida – Dicotyledons
Subclass	Rosidae
Order	Fabales
Family	Fabaceae/Leguminosae – Pea family

Genus	*Senegalia* – Guajillo
Species	*Senegalia berlandieri* or *Acacia berlandieri* Benth - Guajillo

The antinutritional compounds that have been involved in reducing the use of tree and shrub forages comprise glycosides, nonprotein amino acids, polyphenolics, phytohemaglutinins, triter penes, oxalic acid and alkaloids (Table 16.2). *Senegalia berlandieri* comprises a great number of several alkaloids, the most abundant of those are tyramine, phenethylamine and N-methylphenethylamine. The same type of alkaloids were also reported in *Vachellia rigidula*.

Table 16.2. Total alkaloids found in dried leaves

Beta-methyl-phenethylamine	It is a drug stimulant . It is found in many *Acacia* species, notably in *Senegalia berlandieri*.
Catechin	It is a antioxidant natural phenol. The catechin term is frequently used to refer to the related of flavonoid compounds.
Fisetin	It is a flavonol and in many plants is used as coloring agent.
Hordenine	It is a phenylethylamine alkaloid with antibacterial and antibiotic properties. It also motivates the discharge of norepinephrine in mammals.
Phenethylamine	Some studies have found that functions as a neurotransmitter in the central nervous system of mammals
Quercetin	It is a flavonol used as a component in drinks or feedstuffs.
Tyramine	It functions as catecholamine liberating agent. Particularly, however, it is not able to cross the blood-brain-barrier, occasioning in just simply non- psychoactive peripheral sympathomimetic effects.

Data obtained from Beverly *et al.* (1997)

Nutritional value

Chemical analyses may recommend that *S. berlandieri* has high levels of crude protein (CP). However, could be confusing due to a significant quantity of N in the plant is in nonprotein nitrogen (NPN) form. Moreover, N in *S. berlandieri* is poorly digested due to the binding that happens between condensed tannins and protein and between and NDF components during ruminal digestion.

Because the abundance of N in their foliage, *S. berlandieri* is considered as a N complement for range small ruminants. Guajillo and many other indigenous legumes to the flora of southern Texas, USA and northeastern Mexico, may be considered to have high levels of N in their leaves because it fix atmospheric N through symbiosis that is carried out by fungi rhizobium at its roots. Guajillo may be positioned as a non-conventional significant forage for livestock that grows in these areas under extensive management systems due to its high level of N in the foliage. Thus, N contents are high compared with other nutrients, may be revealed by the requirement for N as a part of nucleic acids, chlorophyll, proteins, and other minor components of plants. A great quantities of CP in plants are from enzymatic origin, of which about 50 percent is from ribulose 1,5-diphosphate-carboxylase. Generally, N compounds in grasses and legumes are comparable in their amino acid composition, thus, both offer a balanced feed for ruminants. The amino acids aspartic acid, arginine and glutamic acid are the main N constituents of grasses and legumes. Important quantities of lysine, alanine and glycine are also present.

In a study, the seasonal variation of CP was determined in *S berlandieri* collected in Cienega de Flores, county of the state of Nuevo Leon, Mexico (Table 7.3). It was reported that during the summer had the highest CP content, but in winter was lower, with an annual average of 22%. Apparently, this value is high enough to meet the average protein requirements for small range ruminants. Similarly, it was found that the Guajillo, collected in Linares, NL, Mexico, varied among seasons with a mean value of 19% (Table 7.4). In other study, it was reported also values of 22% when it was collected in Anahuac, Nuevo León in summer (Table 7.5). The CP of alfalfa (*Medicago sativa*) is lower than the Guajillo.

The CP content of Guajillo collected in Texas, USA varied seasonally from a range of 13 to 27.7%. Other study reported lower values of digestible CP due to the high content of condensed tannins (CT). These facts suggest that CT or lignin or insoluble nitrogen in NDF lower the CP quality and quantity for browsers. Condensed tannins may defend shrubs against tissue removal by herbivory, although their efficacy in deterring herbivory by deer has been questioned. Leaf tissues of mature plants of Guajillo contain protein-precipitating tannins and other secondary compounds. Secondary compounds such as tannins and phenols and may defend plants against herbivores in environments where growth is restricted by low soil fertility. Moreover, fertilization may increase palatability of shrubs and increase concentrations of nutrients, making the plants more vulnerable to herbivores. According with this, in four *in vivo* metabolism trials with each mixed diet of 0%, 25%, 50%, and 75% Guajillo by male white-tailed deer, reported that N balance and digestibility diminished with augmented the amount of Guajillo in diets. Urinary glucuronic acid excretion reduced linearly with augmented the quantity Guajillo in diets. In addition, N requirements for body growth and antler development were met by diets containing <60% Guajillo, however, energy needs for maintenance and antler growth were come across with diets having <20% Guajillo. Thus, amounts of Guajillo <20% in diets may sustenance the maintenance needs of white-tailed deer.

Table 16.3. Seasonal dynamic of chemical composition (% dry matter) and molar proportions of volatile fatty acids (mol/100 mol) in leaves of *Senegalia berlandieri*, collected at Cienega de Flores county of the State of Nuevo Leon, Mexico

Item[1]	Seasons				Annual
	Winter	Spring	Summer	Autumn	means
Organic matter	89	93	91	90	91
Ash	11	7	9	10	9
Crude protein	19	20	26	21	22
Degradable protein	6	7	10	8	8
Escape protein	13	13	16	13	15
Neutral detergent fiber	33	36	46	30	36
Cellular content	67	64	54	60	64
ISNDF	7	10	9	7	8
Acid detergent fiber	25	28	37	22	28
Cellulose	8	11	11	6	9
Hemicellulose	8	8	9	8	8
Acid detergent lignin	15	16	16	15	16
Condensed tannins	17	33	24	19	23
Acetic acid	71	74	72	69	72

Propionic acid	19	14	17	18	17
Butyric acid	10	11	10	9	10

ISNDF = insoluble neutral detergent fiber

Data were obtained from Ramírez *et al.* (2000a); Ramírez *et al.* (2000b); Ramírez *et al.* (2000c).

Table 16.4. Chemical composition (%, dry matter) *Senegalia berlandieri* collected during Summer fall and winter in Linares county of the state of Nuevo Leon, Mexico

Concept[1]	Seasons			
	Summer	Fall	Winter	Mean
Organic matter	93	94	94	94
Ash	7	6	6	7
Crude protein	20	18	19	19
Neutral detergent fiber	43	38	38	40
Acid detergent fiber	35	29	32	32
Cellulose	12	10	11	11
Hemicellulose	8	9	6	8
Acid detergent lignin	23	19	20	22
Condensed tannins	10	15	10	12

Data obtained from Ramirez *et al.* (1999).

Ruminants may obtain benefits from several forms of NPN due to rumen icroorganisms can use this type of N to synthetize amino acids that may be absorbed by the ruminant. Otherwise, NPN may be absorbed, metabolized, and excreted without any advantage to the ruminant. For instance, nitrogen-containing chemicals in Guajillo cause toxicities in animals that consume it. Moreover, deer and goats fed augmenting quantities of Guajillo excreted more detoxification conjugates, suggesting guajillo may have compounds that are metabolized through secondary paths. The majority of the N in the form of amino acids, which are absorbed, will be excreted in the urine as ammonium or urea. Nevertheless, a small concentration of the N in amino acids may be excreted in the urine as other metabolic by-products that is describe as uncharacterized N. Nonprotein N that is not used by rumen microbes to produce amino acids and that is absorbed will be detoxified and excreted in the urine as uncharacterized N.

It seems that during summer the NDF content in Guajillo was higher than in other seasons of the year (Tables 16.3 and 16.4). The annual mean was lower than other native legumes, grasses or *Medicago sativa* hay (Table 16.5). It has been determined that forages with low values of NDF may be ingested in larger quantities by small ruminants due to this animals have lower holding rates in the rumen. The lingocellulose in plants represents the acid detergent fiber (ADF) was also inferior, beside with lignin, cellulose and hemicellulose (Tables 16.3, 16.4 and 16.5). Condensed tannins (CT) concentrations varied among seasons and were higher in spring (Table 16.3). However, the NDF content reported in other study carried out by in Guajillo plants collect in Uvalde, Texas, USA, reported higher values during fall than in other seasons (Table 16.6).

Table 16.5. Chemical composition (%, dry matter), molar proportions (mol/100 mol) of volatile fatty acids and *in situ* digestion characteristics (%, dry matter) in *Medicago sativa* hay and leaves of *Senegalia berlandieri* collected in Anahuac county of the state of Nuevo Leon, Mexico

Concept	Medicago sativa	Senegalia berlandieri
Ash	11	7
Insoluble ash	0.01	0.1
Crude protein	18	22
Neutral detergent fiber	36	41
Acid detergent fiber	31	22
Acid detergent lignin	6	4
Cellulose	23	18
Hemicellulose	10	19
Condensed tannins	0.1	11
Acetic acid	69	72
Propionic acid	23	18
Butyric acid	8	10
In situ digestibility parameters		
Dry matter soluble (a), %	35	36
Dry matter degradable (b), %	40	43
Effective degradability of dry matter, 2 %/ hora	69	58
Crude protein soluble (a), %	37	35

Crude protein degradable (b), %	51	43
Effective degradability of crude protein, 2 %/hora	82	57

Data obtained from Ramirez-Lozano and Garcia-Castillo (1996)

Table 16.6. Seasonal variation and digestibility of nutrients in leaves of *Senegalia berlandieri* collected in Uvalde, Texas, EUA

Concept	Seasons				Annual
	Winter	Spring	Summer	Fall	mean
Chemical composition					
Gross energy, Mcal/kg	5	5	5	5	5
Crude protein, %	18	17	18	20	18
Ash, %	7	7	6	5	6
Neutral detergent fiber, %	51	54	59	57	55
Acid detergent fiber, %	28	29	34	27	31
Cellulose, %	18	19	20	16	18
Hemicellulose, %	22	25	26	30	26
Lignin, %	11	10	13	12	12
Condensed tannins %	10	11	12	9	11
N in neutral detergent fiber	49	52	54	49	44
IVDDM, %	35	35	36	42	37
In vivo digestion coefficients %					
Dry matter	37	38	46	48	42
Ash	24	25	17	25	23
Cellular content	74	75	75	83	77
Neutral detergent fiber	3	7	18	23	13
Cellulose	9	6	14	1	8
Hemicellulose	59	62	68	71	65
N in neutral detergent fiber neutro	20	17	17	33	22
Condensed tannins	83	84	88	93	87

[1]Dry matter; IVDDM = in vitro digestibility of dry matter; IVDC = *in vivo* digestibility coefficients. Data obtained from Barnes (1988)

In general, the molar proportions of *in vitro* (Tables 16.3) and *in vivo* (Table 16.5) volatile fatty acids (VFA) contained in *S. berlandieri* were very similar among seasons. Other studies have reported equivalent findings. However, the acetic and propionic acids contained in *M. sativa* hay (Table 16.5), were higher and lower, respectively than in the leaves of *S. berlandieri*. This fact could mean that Guajillo is a food with lower energy content than *M. sativa* due to propionic acid is a precursor glucose.

The high CT content in leaves of *S. berlandieri* may negatively influenced the *in situ* (Table 16.5), *in vivo* (Table 16.6) and predicted (Table 16.7) digestibility coefficients when compared to the *M. sativa* hay. It has been established that CT negatively affected the digestion, and diminished voluntary intake caused mainly by the astringency that provoked the reduction of palatability. The astringency is a sensation caused by the formation of compounds of tannins and salivary glycoproteins. In addition, astringency may increment the saliva production that reduces the palatability. However, tannins may joint dietary proteins in the rumen and by this manner protect the proteins from the action microbial enzymes. These tannin-protein compounds are pH unstable at the abomasum, making the dietary proteins available for digestion and absorption at the lower intestinal track,

In this region, during winter and spring, low rain and low temperatures characterize the environmental conditions. These facts may negatively affected the predicted dry matter digestibility (Table 16.7). However, DE, ME and DMI do not varied among seasons and high enough to gain weight of growing range small ruminants consuming Guajillo foliage.

Table 16.7. Seasonal variation of predicted dry mater digestibility (%), digestible energy (Mcal/kg), metabolizable energy (Mcal/kg) and dry matter intake (g/kg BW/day) of the leaves of *Senegalia berlandieri* collected in northeastern Mexico

Concept	Seasons				Annual
	Winter	Spring	Summer	Fall	mean
Dry matter digestibility	75	67	71	74	72
Digestible energy	3.5	3.2	3.3	3.4	3.3
Metabolizable energy	2.8	2.6	2.7	2.8.	2.7
Dry matter intake	83	83	83	84	83

Data obtained from Ramírez Lozano (2004).

Table 16.8. Seasonal variations of minerals (dry matter) content in leaves of *Senegalia berlandieri* collected at Cienega de Flores county of the State of Nuevo Leon, Mexico

Minerals[1]	Seasons				Annual
	Winter	Spring	Summer	Fall	mean
Macrominerals, g/kg					
Ca	7	7	13	16	11
K	6	9	9	8	8
Mg	4	6	7	7	6
Na	0.3	0.2	0.4	0.4	0.4
P	1	1	1	1	1
Microminerals, mg/kg					
Cu	5	5	4	4	5
Fe	108	92	128	161	123
Mn	39	24	24	21	27
Zn	17	20	19	12	17

Because of minerals are required for all vital processes of the ruminant animal, any excess of deficiency in soils and forages are the responsible of the low productivity and reproductive problems in grazing ruminants. In Table 7.8 are listed the seasonal concentrations of macro and microminerals contained in leaves of *S. berlandieri* collected in northeastern, México. It seems that, in all seasons, the Ca, K, Mg and Fe contents (Tables 16.8 and 16.9) were in enough amounts to meet the requirements of adult ruminants (Table 16.10); however, the Na, P, Mn and Zn were deficient to meet the metabolic requirements of adult ruminants. However, these affirmations have to be taken with caution due to the high CT concentrations in leaves of Guajillo can limit the digestibility and absorption of minerals by ruminants the may consume its foliage.

It appears that, herbivores like to consume minerals when ingested substantial amounts of plant secondary compounds. In a study, reported that feeding mousses with tannins suffered deficiencies of Na, and when the mineral was supplemented in their diets the chronic and acute deficiencies were reduced. It was also explained that minerals may serve as neutralizers of organic acids that are produced during ingestion of secondary compounds. Therefore, herbivores may desire minerals during periods of high consumption of secondary compounds with the purpose to neutralize the effect of organic acids.

Table 16.9. Seasonal content of minerals (dry matter) in leaves of *Senegalia berlandieri* collected in Uvalde, Texas, USA

Concept	Seasons				Annual
	Winter	Spring	Summer	Fall	mean
Macrominerals, g/kg					
Ca	16	18	15	9	14
K	14	13	11	13	13
Mg	3	3	2	2	3
Na	3	3	3	2	3
P	2	2	2	3	2
Microminerals, mg/kg					
Cu	7	7	8	6	7
Fe	175	170	214	112	168
Zn	16	21	29	29	24

Data obtained from Barnes (1988).

It has been demonstrated that because of the low availability of fobs during winter and spring seasons in rangelands of northeastern, Mexico and southern, Texas, USA, the diets of white-tailed deer are mainly composed by tree and shrub foliage with high condensed tannin content. Under these circumstances, deer consume minerals that are used as neutralizers or as precursors to form conjugated compounds that use for detoxification of secondary compounds contained in browse plants.

With the purpose to determine the importance of secondary compounds in the mineral intake, four male white-tailed deer were fed with leaves of *S. berlandieri* in amounts of 0, 25, 50, 75 y 100%. It was found that concentrations of Ca, P y Na diminished, as *S. berlandieri* was incremented in the diets; however, Mg content was similar among diets. Losses of Ca, P and Mn were in the feces; however, Na was excreted via urine. The intake rates of Ca, Mg y Na in diets of 100% *S. berlandieri* exceeded the metabolic requirements of deer. It was concluded that P must be supplemented during periods of low rainfall and high *S. berlandieri* consumption in order to avoid deficiencies of P during female reproduction.

Table 16.10. Mineral requirements of different type of ruminants

Minerales	Beef cattle growing and finishing	Adult sheep	Adult white-tailed deer and goats
Macrominerals, g/kg			
Ca	1.9 - 7.3	2.0 - 8.2	1.3 - 3.3
K	6.0	5.0 - 8.0	1.8 - 2.5
Mn	1.0	1.2 - 1.8	0.8 - 2.5
Na	0.6 - 0.8	0.9 - 1.8	0.6 - 1.0
P	1.2 - 7.4	1.6 - 3.8	1.6 - 3.8
Microminerals, mg/kg			
Cu	10	7 - 11	8 - 10
Fe	50	30 - 50	30 - 40
Mn	20	20 - 40	30 - 40
Zn	30	20 - 33	40 - 50

CHAPTER 17

Nutrition of *Senegalia greggii* Benth.

Introduction

Senegalia wrightii known as cat's claw belongs to the family Fabaceae (Leguminosae), whose synonymy is *Acacia greggii*. The green leaves are used as fodder for domestic ruminants for their high protein content. Butterflies, bees and other insects eat its flowers. White-tailed deer and other small mammals such as rabbits, also consume its foliage. Some mammals and small birds use the plant as a refuge and protection. The flowers are creamy - yellow, oblong yellow puff, sometimes-dull colored; its fragrant bloom covers April to October (spring, summer, autumn). The fruit is a reddish legume often becomes narrow between the seeds. The pods are broad, flat, twist and curl as corn chips when dried. Other collective usages are xeriscape, landscape, as a decorative plant, or border plant. The taxonomic characteristics are listed in Table 17.1.

Senegalia greggii

Table 17.1. Taxonomic characteristics of *Senegalia greggii*

Kingdom	Plantae – Plants
Sub kingdom	Tracheobionta – Vascular plants
Super division	Spermatophyta – Plants with seeds
Division	Magnoliophyta – Plants with flowers
Class	Magnoliopsida – Dicotyledonous
Subclass	Rosidae –
Order	Fabales –
Family	Fabaceae – Leguminosae
Genus	*Senegalia*
Species	*greggii* Hill. – cat claw

One of the main qualities of *Senegalia greggii* is their capacity to prosper on unproductive soils. It is indispensable that valuations be completed of the protagonist the symbiotic rhizobia and ecto and endo-mycorrhizas performance in the progress and improvement of these plants.

Nutritional value

On Tables 17.2., 17.3 and 17.4 are listed the nutritional characteristics of *Senegalia greggii*. The leaves are rich in crude protein, low neutral detergent fiber and its components (cellulose, hemicellulose and lignin). Low levels of condensed tannins. The volatile fatty acids concentrations are similar among seasons being higher the acetic acid than propionic or butyric acids. The predicted dry matter digestibility was higher in fall than other seasons. Similar tendency were determined in predicted digestible and metabolizable energy. The predicted dry matter intake was similar among seasons of the year. It appears that except of Na, al essential minerals listed in Table 17.4 are in sufficient amount to meet metabolic requirements of adult range ruminants.

Table. 17.2. Seasonal variations of the chemical composition (%; dry matter) and molar proportions of the volatile fatty acids (Mol/100 Mol, dry matter) in leaves of *Senegalia greggii* collected in northeastern Mexico

Concept	Seasons				Annual
	Winter	Spring	Summer	Fall	mean
Organic matter	92	91	90	91	91
Ash	8	9	10	10	9
Crude protein	23	24	21	19	22
Degradable protein	14	14	14	13	14
Undegraded protein	9	10	7	6	8
Neutral detergent fiber	50	42	44	39	44
Insoluble neutral detergent fiber	14	14	14	14	14
Acid detergent fiber	34	32	31	27	31
Cellulose	17	20	19	14	17
Hemicellulose	17	10	13	12	13
Lignin	17	13	11	12	13
Condensed tannins	0.6	0.1	0.4	1	1
Acetic acid	68	70	68	69	69
Propionic acid	27	24	27	26	26
Butyric acid	5	6	5	5	5

Table 17.3. Seasonal variation of predicted dry mater digestibility (%), digestible energy (Mcal/kg), metabolizable energy (Mcal/kg) and dry matter intake (g/kg BW/day) of the leaves of *Senegalia greggii* collected in northeastern Mexico

Concept	Seasons				Annual
	Winter	Spring	Summer	Fall	mean
Dry matter digestibility	59	61	61	64	59
Digestible energy	2.8	2.9	2.9	3.0	2.8
Metabolizable energy	2.3	2.4	2.4	2.5	2.3
Dry matter intake	82	83	83	83	82

Table 17.3. Seasonal content of macro (g/kg) and microminerals (mg/kg) in the leaves of *Senegalia greggii* collected in northeastern Mexico

Concept	Seasons				Annual
	Winter	Spring	Summer	Fall	mean
Macrominerals					
Ca	30	31	32	35	32
K	18	15	15	11	15
Mg	4	5	4	3	4
Na	<1	<1	<1	<1	<1
P	2	2	2	2	2
Microminerals					
Cu	23	16	15	20	19
Fe	377	219	283	256	283
Mn	204	132	135	138	152
Zn	84	55	47	41	57

CHAPTER 18

Nutrition of *Senna wislizeni* (A.Gray) Irwin & Barneby

Introduction

The species is usually called Shrubby Senna. This is a <u>perennial</u>, <u>deciduous</u> plant. It is native to <u>Chihuahua</u> and <u>Hidalgo</u>, Mexico and <u>Texas</u>, <u>New Mexico</u>, and <u>Arizona</u> in the USA. It is a desert species with good warmth and dry acceptance. Species of this genus may origin toxic and lethal disease in humans. The sensitivity to a toxin differs with the age of individuals, weight, corporeal status, and specific vulnerability. Kids are more susceptible due to their interest and short size. Poisonousness may vary in plants depending to the season of the year, the diverse parts of the plant species, and its period of growing; and plants may absorb poisonous compounds. It lives on slight water, but gas better growth and better bloom display with few more water. It is a shrubby of winter deciduous.

Senna wislizeni

Table 18.1. Taxonomic characteristics of *Senna wislizeni*

Rank	Scientific Name and Common Name
Kingdom	Plantae – Plants
Subkingdom	Tracheobionta – Vascular plants
Superdivision	Spermatophyta – Seed plants
Division	Magnoliophyta – Flowering plants
Class	Magnoliopsida – Dicotyledons
Subclass	Rosidae
Order	Fabales
Family	Fabaceae/Leguminosae – Pea family
Genus	*Senna* Wight & Arn.
Species	*Senna wislizeni* (A. Gray) Irwin & Barneby – Shrubby senna

Nutritional value

In Tables 18.1, 18.2 and 18.3 are listed the nutritional characteristics of leaves of *Senna wislizeni*, consumed by range ruminates and collected at the north region of Durango State, Mexico. It has medium crude protein content, low neutral detergent fiber. The cellulose content is much higher than hemicellulose or lignin. It has very low values of condensed tannins and ether extract, the acetic acid is very low; however, is much higher than propionic or butyric, isobutyric, isovaleric valeric acids. Except of Na and Zn, other minerals were sufficient to fulfill the adult small ruminant metabolic requirements. The *in vitro* gas production was low along with the bypass protein, microbial protein and metabolizable protein. The potential dry matter digestibility was also medium; however, the potential digestible energy, potential metabolizable energy and potential dry matter intake values were sufficient for the needs of range ruminant consuming leaves of *Senna wislizeni*.

Table 18.2 Chemical composition (%) and *in vitro* volatile fatty acids (mM) of leaves from *Senna wislizeni* collected in the state of Durango México

Concept	Dry matter basis
Ash	10
Crude protein	10
Neutral detergent fiber	49
Acid detergent fiber	33
Lignin	8
Hemicellulose	16
Cellulose	25
Condensed tannins	1
Ether extract	3
Acetic	11
Propionic	2
Butyric	1
Isobutyric	0.2
Isovaleric	0.2
Valeric	0.2
Total	15
Acetic:Propionic	5

Table 18.3. Mineral composition of macro (g/kg, dry matter) and microminerals (mg/kg, dry matter) of leaves from *Senna wislizeni* collected in the state of Durango México

Concept	Minerals
Macrominerals	
Ca	11
K	11
Mg	3
Na	1
P	3

Microminerals

Cu	9
Fe	239
Mn	43
Zn	26

Table 18.4. *In vitro* gas production in 24 hours ($GasP_{24h}$/ mL 200 mg, dry matter), bypass protein (%, dry matter), microbial protein (%, dry matter) and metabolizable protein (%, dry matter) in leaves from *Senna wislizeni* collected in north of the state of Durango México

Concept	Dry matter
In vitro gas production	40
Bypass protein	1
Microbial protein	5
Metabolizable protein	6
Potential dry matter digestibility	58
Potential digestible energy, Mcal/kg	2.8
Potential metabolizable energy, Mcal/kg	2.3
Potential dry matter intake, g/kg/day	84

CHAPTER 19

Nutrition of *Vachellia constricta* (Benth.) Seigler & Ebinger

Introduction

This species is commonly known as the whitethorn acacia. It is a shrub native to northern Mexico and the Southwestern United States. It belongs to the Fabaceae family. Whitethorn acacia is has not being considered as favorite plant species for cattle. However, it was suggest that plant is grazed by cattle when no better feed source is accessible. Large mammals such as desert mule deer do not browse extensively on *V. constricta*, however, the plant sometime makes up a small quantity of their diets. It has been reported that white-tailed deer browse the leaves and pods. In a study, it was found the diets of desert mule deer were 0.4% *V. constricta* during winter. Similarly, in other desert mule deer diet studies, it was reported low use of the plant (1-5%) with 0% during spring. This low usage of whitethorn acacia by mule deer suggests that the plant can be used when species that are more palatable are rare.

Vachellia constricta

Table 19.1 Taxonomic characteristics of *Vachellia constricta*

Rank	Scientific Name and Common Name
Kingdom	Plantae – Plants
Subkingdom	Tracheobionta – Vascular plants
Superdivision	Spermatophyta – Seed plants
Division	Magnoliophyta – Flowering plants
Class	Magnoliopsida – Dicotyledons
Subclass	Rosidae
Order	Fabales
Family	Fabaceae/Leguminosae – Pea family
Genus	Vachellia Wight & Arn. – acacia
Species	*Vachellia constricta* (Benth.) Seigler & Ebinger – whitethorn acacia

Nutritional value

In Tables 19.2, 19.3 and 19.4 are listed the nutritional characteristics of leaves of *Vachellia constricta*, consumed by range ruminates and collected at the north region of Durango State, Mexico. It has high crude protein content, low neutral detergent fiber similar values of hemicellulose adn cellulose, very low values of condensed tannins and ether extract, the acetic acid is much higher than propionic or butyric, isobutyric, isovaleric valeric acids. Except of Mg, Na and Zn, other minerals were good enough to fulfill the adult small ruminant metabolic requirements. The *in vitro* gas production was low along with the bypass protein, microbial protein and metabolizable protein. The potential dry matter digestibility was also medium; however, the potential digestible energy, potential metabolizable energy and potential dry matter intake values were sufficient for the needs of range ruminant consuming leaves of *Vachellia costricta*.

Table 19.2. Chemical composition (%) and *in vitro* volatile fatty acids (mM) of leaves from *Vachellia constricta* collected in the state of Durango México

Concept	Dry matter basis
Ash	7
Crude protein	17
Neutral detergent fiber	35
Acid detergent fiber	22
Lignin	9
Hemicellulose	13
Cellulose	13
Condensed tannins	3
Ether extract	2
Acetic	12
Propionic	3
Butyric	2
Isobutyric	0.4
Isovaleric	0.4
Valeric	0.5
Total	18
Acetic:Propionic	4

Table 19.3. Mineral composition of macro (g/kg, dry matter) and microminerals (mg/kg, dry matter) of leaves from *Vachellia constricta* collected in the state of Durango México

Concept	Minerals
Macrominerals	
Ca	21
K	9
Mg	1
Na	1
P	1.9

Microminerals

Cu	9
Fe	60
Mn	129
Zn	18

Table 19.4. *In vitro* gas production in 24 hours (GasP$_{24h}$/ mL 200 mg, dry matter), bypass protein (%, dry matter), microbial protein (%, dry matter) and metabolizable protein (%, dry matter) in leaves from *Vachellia constricta* collected in north of the state of Durango México

Concept	Dry matter
In vitro gas production	38
Bypass protein	2
Microbial protein	7
Metabolizable protein	10
Potential dry matter digestibility	69
Potential digestible energy, Mcal/kg	3.2
Potential metabolizable energy, Mcal/kg	2.6
Potential dry matter intake, g/kg/day	83

CHAPTER 20

Nutrition of *Vachellia farnesiana* (L.) Wight et Arn.

Introduction

Vachellia farnesiana, also known as *Acacia farnesiana*, its common names are Sweet Acacia, Perfume Acacia and Huisache. This plant is considered as a useful species for windbreaks, reforestation of dry forests and grasslands degraded areas and to stabilize shifting sands in degraded semiarid regions. The leaves, pods, stems and flowers are used as fodder for range small ruminants, especially during the winter. The foliage and bark have an unpleasant odor and is supposed to pass a bad taste to milk. Because of its height, it is necessary to cut the branches (pruning) for maximum utilization by browsing ruminants. However, in some places it is considered an invader because of its ability to colonize pastures and other disturbed habitats. In northeastern, Mexico and southern Texas, USA, its foliage is avidly consumed by domestic and wild ruminants such as white-tailed deer. However, the relative high level of condensed tannins in their leaves, affects consumption and digestibility of nutrients when is part of the diets of sheep and goats. The taxonomic characteristic of Huisache are listed in Table 8.1.

Vachellia farnesiana

Table 20.1. Taxonomic classification of *Vachellia farnesiana*

Kingdom	Plantae – Plants
Sub kingdom	Tracheobionta – Vascular plants
Super division	Spermatophyta – Plants with seeds
Division	Magnoliophyta – Plants with flowers
Class	Magnoliopsida – Dicotyledonous
Subclass	Rosidae –
Order	Fabales –
Family	Fabaceae – Leguminosae
Genus	*Vachellia* or *Acacia*
Species	*Farnesiana* (L.) Wild - Huisache

Nutritional value

The crude protein (CP) content of the leaves of *Vachellia farnesiana* was very similar among seasons (Table 20.2). The CP levels throughout the year were sufficiently high to meet the requirements of any ruminant species for growth and reproduction including white-tailed deer (Table

20.3). These results are similar to those found in evaluated plants collected in southern Texas, USA. They reported values ranging from 27, 26 and 21% in sprouts in seed and mature leaves, respectively. In another study, it was reported that the CP content of the leaves collected in summer, autumn and winter, in Linares county of the state of Nuevo Leon, Mexico, were lower with values of 19, 18 and 16 %, respectively (Table 20.4) or in leaves collected in Anahuac county of the state of Nuevo Leon, Mexico (Table 20.5). It seems that CP can vary depending of state of maturity. It has been reported in a comparative study that mature leaves have lower CP content (21%) than immature leaves (28%). The high levels of undegraded protein (Table 20.2) in leaves of V. farnesiana may be due to the condensed tannins content that protected a great proportion of the leaf protein; this protein may be available for digestion in the intestines.

Lower and very similar neutral detergent fiber (NDF) and acid detergent fiber (ADF) contents, in all seasons, are presented in Tables 20.2, 20.3 and 20.5; even lower than Medicago sativa hay. The low fiber and consequent high cell content make this plant with high nutritional value compared to grasses. The hemicellulose was higher than cellulose content. Other studies have reported that leaves from native tropical legumes had lower hemicellulose content than cellulose. The lignin content was very similar among seasons. In addition, lignin was higher than that of M. sativa hay (Table 20.5). The condensed tannin content was low in all seasons and in all collected plants.

The volatile fatty acid (VFA) content was very similar among seasons (Tables 20.2 and 20.5); even similar than M. sativa hay. The digestion coefficients, energy values and dry matter intake (Tables 20.5 and 20.6) were also very similar among seasons. The Ca, K, Mg, Fe, Mn and Zn concentrations were sufficient to meet metabolic requirements of adult range ruminants; however, Na, P and Cu were lower (Table 20.8).

Table 20.2. Seasonal variation of the chemical composition (% dry matter), and molar proportions of volatile fatty acids (Mol/100 Mol) of leaves of *Vachellia farnesiana* collected in Cienega de Flores county of the state of Nuevo Leon, Mexico

Concept	Seasons				Annual
	Winter	Spring	Summer	Fall	mean
Organic matter	91	91	93	95	93
Ash	9	9	7	5	7
Crude protein	22	21	22	20	21
Degradable protein	14	12	11	9	12
Undegraded protein	8	9	11	11	9
Neutral detergent fiber	36	38	35	38	37
Insoluble neutral detergent fiber	7	5	6	7	6
Cellular content	64	62	65	62	63
Acid detergent fiber	23	24	22	24	23
Cellulose	8	9	7	9	8
Hemicellulose	13	14	13	14	14
Lignin	14	14	13	15	14
Condensed tannins	2	2	5	5	2
Acetic acid	70	71	70	69	70
Propionic acid	26	23	25	25	25
Butyric acid	4	6	6	6	6

Table 20.3. Requirements of crude protein (CP, %), total digestible nutrients (TDN, %), digestible energy (DE, Mcal/kg) and metabolizable energy (ME, Mcal/kg) of beef cattle, sheep, goats and white-tailed deer.

Animals	Requirements											
	Maintenance				Growing				Final of gestation			
	CP	TDN	DE	ME	CP	TDN	DE	ME	CP	TDN	DE	ME
Beef cows[1]	8.5	55	2.3	2.0	10	64	2.7	2.3	16	52	2.2	1.9
Adult sheep[2]	7.5	41	1.8	1.5	10	60	2.6	2.2	17	108	4.5	3.9
Adult goats[3]	7.0	46	2.0	1.7	9	55	2.3	2.0	13	84	3.5	3.0
Adult white-tailed deer[4]	7.0	51	2.2	1.9	15	60	2.6	2.2	18	59	2.5	2.1

[1]Beef cows of 400 kg with a daily weight gain of 1.0 kg.

[2]Adult sheep with a daily weight gain of 50 g.
[3]Adult goats with a daily weight gain of 50 g.
[4]Adult white-tailed deer with a daily weight gain of 50 g.
Data obtained from NRC, 2000; Kearl, 1981 and Feist, 1998.

Table 20.4. Chemical composition (%, dry matter) of leaves from *Vachellia farnesiana* collected in Linares county of the state of Nuevo Leon, Mexico

Concept	Seasons			
	Summer	Fall	Winter	Mean
Organic matter	92	93	90	92
Ash	8	7	10	8
Crude protein	19	18	16	18
Neutral detergent fiber	36	38	38	37
Acid detergent fiber	27	28	29	28
Cellulose	11	11	12	11
Hemicellulose	8	11	9	9
Lignin	17	15	16	16
Condensed tannins	2	1	1	2

Table 20.5. Chemical composition (%, dry matter), molar proportions of volatile fatty acids (Mol/100 Mol), in *Medicago sativa* hay and *Vachellia farnesiana* leaves collected in Anahuac county of the state of Nuevo Leon, Mexico

Concept	*Medicago sativa* hay	*Acacia farnesiana* leaves
Crude protein	17	20
Neutral detergent fiber	46	40
Acid detergent fiber	33	23
Lignin	3	10
Cellulose	26	23
Hemicellulose	13	15
Condensed tannins	0	2
Acetic acid	69	70
Propionic acid	23	25
Butyric acid	8	6
In vitro dry matter digestibility	61	46

TDN	64	48
Digestible energy, Mcal/kg	2.8	2.1
Metabolizable energy, Mcal/kg	2.4	1.7
Dry matter intake, g/kg BW/day	86	86

Data obtained from Ramirez and Ledezma (1997)

Table 20.6. Seasonal variation of predicted dry mater digestibility (%), digestible energy (Mcal/kg), metabolizable energy (Mcal/kg) and dry matter intake (g/kg BW/day) of the leaves of *Vachellia farnesiana* collected in northeastern Mexico

Concept	Seasons				Annual
	Winter	Spring	Summer	Fall	mean
Dry matter digestibility	74	73	75	72	74
Digestible energy	2.3	2.3	2.3	2.3	2.3
Metabolizable energy	1.9	1.9	1.9	1.9	1.9
Dry matter intake	83	83	83	83	83

Data obtained from Ramirez Lozano (2004).

Table. 20.7. Mineral content of macro (g/kg, dry matter) and microminerals (mg/kg, dry matter) in leaves of *Vachellia farnesiana* in Cienega de Flores county of the state of Nuevo Leon, Mexico

Minerals	Seasons				Annual
	Winter	Spring	Summer	Fall	mean
Macrominerals					
Ca	11	10	12	8	10
K	11	7	11	7	9
Mn	7	9	10	6	8
Na	<1	<1	<1	<1	<1
P	1	1	1	1	1
Microminerals					
Cu	4	4	4	4	4
Fe	105	115	112	108	110
Mg	42	30	29	31	34
Zn	43	33	32	28	30

CHAPTER 21

Nutrition of *Vachellia schaffneri* (S. Watson) F.J. Herm.

Introduction

Vachellia schaffneri commonly known as twisted acacia is a tree native to north Mexico and Texas United States of America. The trees help as feed for range ruminants. Goats and sheep browse the leaves from the plant and consume the beans when accessible at the end of the summer. Livestock use the trees for shade and shelter. The leaves contain some chemical compounds such as Phenethylamine, Beta-methyl-phenethylamine, Tyramine and Hordenine. The foliage and seeds of the species have a protein content of about 11.6%. White-tailed deer browse it, and javelina, feral hogs and some birds eat the fruit. The taxonomic characteristic of *Vachellia schaffneri* are listed in Table 21.1

Table 21.1. Taxonomic characteristics of *Vachellia schaffneri*

Rank	Scientific Name and Common Name
Kingdom	Plantae – Plants
Subkingdom	Viridiplantae
Superdivision	Embryophyta
Division	Tracheophyta – vascular plants, tracheophytes
Class	Magnoliopsida – Dicotyledons
Subclass	Rosidae
Order	Fabales
Family	Fabaceae/Leguminosae – Pea family
Genus	Vachellia Wight & Arn.
Species	*Vachellia schaffneri* (S. Watson) Seigler & Ebinger – twisted acacia

Nutritional value

In Tables 21.2, 21.3 and 21.4 are listed the nutritional characteristics of leaves of *Vachellia schaffneri*, consumed by range ruminates and collected at the north region of Durango State, Mexico. It has high crude protein content, low neutral detergent fiber similar values of hemicellulose and cellulose, very low values of condensed tannins and ether extract, the acetic acid is very low; however, is much higher than propionic or butyric, isobutyric, isovaleric valeric acids. Except of Na and Cu, other minerals were good enough to satisfy the adult small ruminant metabolic requirements. The *in vitro* gas production was low along with the bypass protein, microbial protein and metabolizable protein. The potential dry matter digestibility was also medium; however, the potential digestible energy, potential metabolizable energy and potential dry matter intake values were sufficient for the needs of range ruminant consuming leaves of *Vachellia schaffneri*.

Table 21.2. Chemical composition (%) and *in vitro* volatile fatty acids (mM) of leaves from *Vachellia schaffneri* collected in the state of Durango México

Concept	Dry matter basis
Ash	7
Crude protein	16
Neutral detergent fiber	57
Acid detergent fiber	34
Lignin	11
Hemicellulose	23
Cellulose	24
Condensed tannins	4
Ether extract	1
Acetic	8
Propionic	1
Butyric	0.9
Isobutyric	0.4
Isovaleric	0.3
Valeric	0.4

Total	11
Acetic:Propionic	7

Table 21.3. Mineral composition of macro (g/kg, dry matter) and microminerals (mg/kg, dry matter) of leaves from *Vachellia schaffneri* collected in the state of Durango México

Concept	Minerals
Macrominerals	
Ca	9
K	10
Mg	2
Na	1
P	3
Microminerals	
Cu	4
Fe	153
Mn	47
Zn	54

Table 21.4. *In vitro* gas production in 24 hours ($GasP_{24h}$/ mL 200 mg, dry matter), bypass protein (%, dry matter), microbial protein (%, dry matter) and metabolizable protein (%, dry matter) in leaves from *Vachellia schaffneri* collected in north of the state of Durango México

Concept	Dry matter
In vitro gas production	14
Bypass protein	8
Microbial protein	2
Metabolizable protein	10
Potential dry matter digestibility	58
Potential digestible energy, Mcal/kg	2.8
Potential metabolizable energy, Mcal/kg	2.3
Potential dry matter intake, g/kg/day	81

CHAPTER 22

Nutrition of *Vachellia rigidula* (Benth.) Seigler & Ebinge

Introduction

Common names: Blackbrush acacia, Blackbrush, Chaparro prieto, Gavia. Belongs to the Legume family (Fabaceae). Its native range stretches from Texas in the United States south to central Mexico. It is spiny, stiff-branched, thicket-forming shrub bearing numerous spikes of yellow flowers. Blackbrush grows from 2 to 5 m. Prolific spikes of pale yellow, fragrant flowers are borne on the numerous stiff, thorny branches. The bark of this shrub is whitish in color. Its semi-evergreen leaves are dark-green, glossy and pinnately compound. It is a small perennial tree. The leaves are deciduous and alternated. The fruit type is a legume, the flowers are in 5 cm spikes and the fruit is brown. The bloom **color is white, yellow and bloom time is in March, April, May and June.**

Vachellia rigidula

A pharmacological study of *Vachellia rigidula* found that the existence of over forty alkaloids, including low amounts of several amphetamines that had previously been reported in the related species *Vachellia berlandieri*. The chemicals reported in the highest amounts were phenethylamine, N-methylphenethylamine, tyramine and N-methyltyramine. Other important phytochemicals found were N,N-dimethyltryptamine, mescaline, and nicotine, even though these were reported in low concentrations.

Vachellia rigidula has been used in weight loss dietary supplements because of its adrenergic amine content. These chemicals are believed to stimulate β-receptors to stimulate lipolysis and metabolic rate and reduce appetite. In addition, is known as a great honey maker and early blooming plant for its native origin. In 2015, 52% of supplements labeled as containing *Vachellia rigidula* were found to contain BMPEA, an amphetamine isomer. Consumers following recommended maximum daily servings would consume a maximum of 94 mg of BMPEA per day. Nevertheless, in 2012, the FDA resolute that BMPEA was not naturally existent in *Vachellia rigidula* foliage.

Nutritional value

Forage from browse species has been reported as the main component in diets of range goats, sheep and white-tailed deer (*Odocoileus virginianus*) in semiarid regions regions. *Vachellia rigidula*, **Leucophyllum frutescens, Parkinsonia texana, Celtis pallida, Porlieria angustifolia** and **Cordia boissieri** were most selected by goats. Moreover, *Vachellia rigidula* **represented about 50% of the annual diet.** These species have foliage throughout the year, with enough contents of crude protein (CP) and dry matter (DM) for facing the demands of small ruminants grazing in different physiological conditions. Those tannins-rich species such as *V. rigidula* have an agronomic advantage of adaptation to biotic and environmental stresses over the non-tannin-containing plants; therefore, the usage of these species for livestock may guarantee animal productivity.

However, shrubs quality can be affected by the maturity of plant, soil management and secondary metabolites, such as condensed tannins, which form different complexes with plant compounds including protein and in some situations, decreasing digestibility. In Table 22.1 the organic matter content of leaves from *Vachellia rigidula* varied in a narrow but significant range throughout the year. In addition, *V. rigidula,* in general, had more CP during winter and summer and low during autumn and spring (Table 22.1). In a study reported that forage from trees and browse plants have higher CP compared to that at the end of the spring and at the beginning of the fall. In other study, established that in many regions of the world, CP content in browse plants is high, and it is relatively constant throughout the year, compared to grass.

Low neutral detergent fiber values, during all seasons, are presented in Table 22.1. It has been established that browse plants with low NDF content have a higher nutritional quality and forages with immature growth have lesser NDF. In a study also reported that NDF in all evaluated shrubs was lesser to that of the *Medicago sativa* hay. This fact may represent an advantage for ruminants that are fed with these browse plants, because forages that contain low fiber could be digested more easily and in more quantity. The highest level in lignin in the browse plants occurred in spring and fall. Most plants had an annual mean values of lignin higher that alfalfa hay. Condensed tannins (Table 5.1) were higher in spring and summer. It seems that the levels of condensed tannins, encountered in *V. rigidula* are toxic for small ruminants. However, goats and white-tailed deer have demonstrated an ability to utilize browse with high content of plant secondary compounds such as condensed tannins. The molar proportions of volatile fatty acids reported in Table 5.1 are comparable to values reported for commercial forages such as *M. sativa* hay.

Table 22.1. Seasonal dynamics of the chemical composition and molar percentage of the volatile fatty acids of nutrients contained in de la *Vachellia rigidula*

Concept[1]	Seasons				Annual
	Winter	Spring	Summer	Fall	mean
			%		
Organic matter	91	95	88	92	93
Ash	9	5	12	8	8
Crude protein	17	16	17	13	17

Degraded protein	5	5	4	3	4
Undegraded protein	12	11	13	10	13
Neutral detergent fiber	39	40	48	53	44
INDF	10	9	13	16	11
Cellular content	61	60	52	47	56
Acid detergent fiber	28	23	36	46	30
Cellulose	16	18	12	13	14
Hemicellulose	11	17	10	7	12
Lignin	12	12	16	17	14
Condensed tannins	15	18	22	22	20
Volatile fatty acids	% molar (moles/100 moles)				
Acetic acid	69	69	69	74	70
Propionic acid	29	29	27	23	28
Butyric acid	2	2	4	5	3

[1]% dry matter; NDF = neutral detergent fiber; INDF = insoluble neutral detergent fiber.

Table 22.2. Seasonal variation of digestion, energy and intake of *Vachellia rigidula*

Concept	Seasons				Annual
	Winter	Spring	Summer	Fall	mean
Dry matter digestibility, %	68	71	61	51	66
Digestible energy, Mcal/kg	2.12	2.23	1.91	1.60	2.06
Metabolizable energy, Mcal/kg	1.74	1.83	1.27	1.31	1.70
Dry matter intake, g/kg BW/day	83	83	82	82	82

Dry matter digestibility was calculated as 83.58 - 0.824*acid detergent fiber + 2.626*nitrogen (Odd *et al.*, 1983).
Digestible energy calculated as 0.27+0.0428*DMD (Fonnesbeck *et al.*, 1984).
Metabolizable energy calculated as 0.821*DE (Khalil *et al.*, 1986).
IDNDF = indigestible neutral detergent fiber calculated as 86.98+1.542*NDF+31.63*Acid detergent fiber (Jancik *et al.*, 2008).
Dry matter intake was calculated as 86.5-0.09*neutral detergent fiber.

Predicted dry matter digestibility (DMD), digestible energy (DE), metabolizable energy (ME) and dry matter intake (DMI) of leaves from *V. rigidula* are listed in Table 5.2. Lower levels of neutral detergent fiber (NDF) and its constituents (cellulose, hemicellulose and lignin) and condensed tannins (CT; Table 5.1) might have influenced the seasonal content of DMD, DE, ME and DMI. In winter and spring when dietary cell wall content were lower, the digestion and intake parameters were higher. Conversely, in summer and fall were lower.

In Table 22.3 are shown the seasonally contents of Ca, P, Mg, K, Zn, Mn, Cu and Fe of leaves of *Vachellia rigidula* collected in northeastern Mexico. It appears that despite seasonal variations leaves of *Vachellia rigidula* had Ca levels that could meet the adult range small ruminant requirements. It has been also reported that leaves from browse plants that grow in semiarid and tropical regions have enough Ca for optimal livestock and wildlife performance. It appears that leaves of *V. rigidula* contains K as much as 10 times the required levels. This fact may become a problem because high K concentrations can interfere with Na retention, absorption and Mg utilization. It seems that in all seasons the Mg concentrations were sufficient to meet adult range small ruminant requirements. It has been reported high Mg in 18 shrubs that grow in Texas, USA.

Table 22.3. Seasonal mineral concentrations in the leaves of *Vachellia rigidula*

Minerales[1]	Seasons				Annual
	Winter	Spring	Summer	Fall	mean
Macro, g/kg					
Ca	10	5	8	8	8
K	9	10	8	5	8
Mg	3	5	4	2	4
Na	0.4	0.3	0.2	0.4	0.4
P	1	1	1	1	1
Micro, mg/kg					
Cu	5	9	6	6	6
Fe	74	112	100	110	99
Mn	15	22	20	21	20
Zn	17	23	21	26	22

[1]Dry matter; Data obtained from: Ramirez *et al.*, 2001.

Apparently, during spring *V. rigidula* had higher Cu concentrations than in other seasons. In other seasons, Cu levels were marginal below to meet adult range small ruminant requirements. Low Cu concentrations are also reported in shrubs from semiarid regions and in tropical legumes. All shrubs, in all seasons, contained Fe levels in substantial amounts to meet adult range small ruminant requirements. It has been reported that high levels of condensed tannins in browse plants can decrease Fe absorption. The shrub *V. rigidula* has high level of condensed tannin content (Table 22.1), thus, high adult range small ruminant intakes of these plants may decrease Fe absorption. However, it is known that browsing animals have salivary enzymes that degrade or bind tannins, thus their presence may not inhibit digestion of plants nor the absorption of plant mineral content. The Mn concentrations were marginal below to meet goat requirements.

It has been reported that with the exception of fruits from *A. berlandieri, Acacia tortuosa* and *P. glandulosa*, all evaluated plants that occur in south Texas, USA, had low levels of Mn to meet the requirements of grazing cattle. In all seasons, Zn content was insufficient to meet range small ruminant requirements. Some shrubs that occur in Texas, USA, had Zn levels that varied seasonally, but only a few of them had levels of Zn to meet domestic livestock and white-tailed deer requirements. In a study was evaluated the nutritional value of *Vachellia rigidula,* a legume species that grow in the Tamaulipan Thornscrub Vegetation, northeastern Mexico. In a first experiment was tested the effects of season on chemical composition as nutritional variable to small ruminants; the second tested the effect of the addition of polyethylene glycol (PEG) on *in vitro* fermentation parameters. It was found that chemical composition was affected by season (Table 5.4). Moreover, addition of PEG increased fermentation parameters and metabolizable energy content (Table 22.5).

Results indicate that chemical composition and fermentation parameters vary according to seasons. The PEG addition increased the fermentation parameters, which indicates that PEG counteracts the detrimental effects of condensed tannins of leaves from *V. rigidula*. Therefore, data suggest that *V. rigidula* could supply necessary nutritional requirements to small ruminants (goat and white-tailed deer) in the Tamaulipan Thornscrub Vegetation of northeastern Mexico.

Table 22.4. Seasonal variation of the chemical composition, *in vitro* true organic matter digestibility (IVTOMD, %) and gross energy losses (GE, % as methane production) of leaves of *Vachellia rigidula* collected in northeastern Nuevo Leon, Mexico.

Concept	NDF	CT	CP	EE	IVTOMD	GE, CH4
Spring	52.8	18.5	15.0	0.6	42.6	7.1
Summer	45.4	20.3	13.6	1.0	46.8	6.9
Autumn	49.6	18.4	15.4	0.8	53.2	6.5
Winter	48.0	18.8	15.2	1.6	58.7	6.2
Mean	49.0	19.0	14.8	1.0	50.3	6.7

NDF = neutral detergent fiber; CT = condensed tannins; CP = crude protein; EE = ether extract

Table 22.5. Effect of polyethylene glycol on gas production (Gas24 h, ml/g dry matter), microbial protein (MP, µmol), partitioning factor (PF) and metabolizable energy (ME, Mcal/kg dry matter) of leaves of *V. rigidula* in northeastern Nuevo Leon, Mexico

Concept	Gas24 h	Rate, h	Lag time, h	MP	PF	ME
Without PEG						
Spring	109	0.04	0.42	6.1	2.8	1.2
Summer	97	0.02	0.40	5.5	3.0	1.9
Autumn	89	0.04	0.72	13.2	3.7	2.2
Winter	105	0.04	0.33	7.0	4.3	1.5
Mean	100	0.04	0.47	8.0	3.5	1.7
With PEG						
Spring	164	0.06	0.14	6.9	4.9	1.6
Summer	163	0.06	0.59	7.1	5.9	2.0
Autumn	157	0.06	0.93	15.2	7.2	2.1
Winter	147	0.06	0.66	6.9	6.5	1.8
Mean	158	0.06	0.58	9.0	6.1	1.9

CHAPTER 23

Nutrition of *Atriplex canescens* (Pursh) Nutt.

Introduction

Atriplex canescens, commonly known as saltbush, four-wing saltbush, and fourwing saltbush, is an evergreen shrub belonging to the family Amaranthaceae that is native to the western of USA. The saltbush is a shrubby species resistant to drought, widespread also in the arid areas of northern Mexico. It prospers in sandy soils or sandy loam, but also grow well in clay loam and clay soils; it is in addition tolerant to moderate to high concentrations of salinity and alkalinity. The use of Atriplex bushes has been in some cases as a way to improve the productivity of rangelands, by replacing native vegetation or has been used to reforest degraded areas. However, in several countries the saltbush is used as a means to feed ruminant grazing during critical times of forage production. The taxonomic characteristic of *Atriplex canescens* are listed in Table 23.1.

Atriplex canescens

Table 23.1. Taxonomic characteristics of *Atriplex canescens*

Rank	Scientific Name and Common Name
Kingdom	Plantae – Plants
Subkingdom	Tracheobionta – Vascular plants
Superdivision	Spermatophyta – Seed plants
Division	Magnoliophyta – Flowering plants
Class	Magnoliopsida – Dicotyledons
Subclass	Caryophyllidae
Order	Caryophyllales
Family	Chenopodiaceae – Goosefoot family
Genus	*Atriplex* L. – saltbush
Species	*Atriplex canescens* (Pursh) Nutt. – fourwing saltbush

Nutritional value

Saltbush has been described as one of the most valuable fodder shrub desert as it has been attributed features such as palatability, nutritional value, accessibility to browsing and abundance in forage production. Saltbush palatability appears to be due to the amount of accumulated salts in leaves. As for its nutritional value, the saltbush has a crude protein content (16-20%) than do comparable with some other species such as alfalfa. The crude fiber content increases with the maturity of the bush, unlike its protein content remains constant during most of the year. As for his mineral content, it has been mentioned that there is an increase in the leaves during winter. The phosphorus content varies depending on the time of year, being the highest values either in the leaves and stems in spring (0.23% in leaves and 0.17% in the stems) to decrease later in the winter (0.10% to 0.05% in leaves and stems, respectively).

Studies of *in vitro* dry matter digestibility (IVDMD) reported values of 30% to 60 in leaves, whereas the stems is decreasing as winter approaches of 30 to 50% in spring and 10% in winter. Digestibility in vivo studies with sheep determined the apparent digestibility of two species of *Atriplex* spp (*A. canescens* and *A. acanthocarpa*); the authors found an average digestibility of the two species of 52.6%, lower than that of alfalfa hay that was 62.5%. In another study, the chemical composition and the *in vitro* digestion in samples of four mature *Atriplex* species (A. *canescens, A. barclayana, A. lentiformis* and A. *nummularia*) irrigated with hypersaline water was reported. Some of these were analyzed without desalting and the other was subjected to desalting process. The results obtained for the different species the crude protein in the first case was between 7.6 and 11.5%, while the IVDMD was 49.0 to 58.8%. Desalinated crude protein samples showed losses and decreased digestibility; It is the most affected A. *nummularia* (22.1% and 50.78%, respectively). It was found that the concentration of the various factors considered anti-nutritional or toxic had no influence to the animals. Therefore, it was concluded that the four species of Atriplex analyzed represent a potential forage for feeding ruminants in the arid and semi-arid areas and the desalting process tested was possible to reduce the salt content in plant up to 98.20%, although it could not avoid the loss of protein.

It is important to consider that even if the saltbush has a good nutritional quality, the botanical composition of the diet of grazing ruminants may vary, depending on the presence of other shrubs, grasses and herbs. The selectivity of the animal can change according to events such as rainfall, temperature, the availability of other fodder on the pasture and animal as needed to fill their requirements for energy and protein. In a study in Chihuahua, Mexico during the dry season (January to June), the saltbush was 65.9% of the diet of grazing cattle. In another study, conducted in northern Zacatecas, Mexico, the botanical composition of the diet of goats grazing in a brush of Larrea-fluerensia-Atriplex consisted of high levels of *Atriplex* spp, 59% on average throughout the year. However, in other study, the saltbush represented less than 20% of the diet, possibly due to low availability of this shrub in the pasture, or by the presence of others that were more selected species by goats. The time of daily grazing influences the amount of forage that the animal can consume. In a study conducted in the winter-spring season in North Yemen (Arabic) with grazing sheep for 30 minutes a day in a meadow of saltbush, the animals did not lose weight than the control group. The author calculated consumption of 70 g/day of dry matter for a period of *Atriplex* grazing seven weeks.

The effect of consumption of *Atriplex* spp by replacing alfalfa on growing kid performance has been reported. The weight gain obtained by providing 50% alfalfa and 50% *Atriplex* was 77 g/day, compared to 146 g/day by providing 100% alfalfa. The authors mentioned that the food energy remained below requirements. This factor probably affected weight gains obtained. Probably the saltbush forage is limited in terms of energy supply, not as to the supply of crude protein. Furthermore, these results indicate that the saltbush can be used as a protein supplement in livestock feed goats managed under an extensive system of free ranging grazing. The authors conclude that food energy remained below requirements. This factor is likely to affect weight gains obtained thus may limit forage saltbush energy intake in the diet, but not the supply of crude protein. Thus, in the arid and semi-arid regions of northeastern Mexico, where range goats traditionally eat, diets based on browse and forage plants, saltbush could be used as good protein supplement.

In Tables 23.2, 23.3 and 23.4 are listed the nutritional characteristics of leaves of *Atriplex canescens*, consumed by range ruminates and collected in summer at the north region of Durango State, Mexico. It has medium crude protein content, low neutral detergent fiber similar and lignin. Values of hemicellulose are higher than cellulose, very low values of condensed tannins and ether extract, the acetic acid is very low; however, is much higher than propionic or butyric, isobutyric, isovaleric valeric acids. Except of Na and Cu, other minerals were good enough to satisfy the adult small ruminant metabolic requirements. The *in vitro* gas production was low along with the bypass protein, microbial protein and metabolizable protein. The potential dry matter digestibility, the potential digestible energy, potential metabolizable energy and potential dry matter intake values were sufficient for the needs of range ruminant consuming leaves of *Atriplex canescens*.

Table 23.2. Chemical composition (%) and volatile fatty acids (mM) of leaves from *Atriplex canescens* collected in the north part of the state of Durango, Mexico

Concept	Dry matter basis
Ash	27
Crude protein	12
Neutral detergent fiber	35

Acid detergent fiber	11
Lignin	3
Hemicellulose	24
Cellulose	8
Condensed tannins	0.3
Ether extract	1
Acetic	7
Propionic	1
Butyric	1
Isobutyric	0.1
Isovaleric	0.1
Valeric	0.4
Total	8.5
Acetic:Propionic	11

Table 23.3. Mineral composition of macro (g/kg, dry matter) and microminerals (mg/kg, dry matter) of leaves from *Atriplex canescens* collected in the state of Durango México

Concept	Minerals
Macrominerals	
Ca	22
K	56
Mg	10
Na	5
P	3
Microminerals	
Cu	12
Fe	112
Mn	66
Zn	71

Table 23.4. *In vitro* gas production in 24 hours (GasP$_{24h}$/ mL 200 mg, dry matter), bypass protein (%, dry matter), microbial protein (%, dry matter) and metabolizable protein (%, dry matter) in leaves from *Atriplex canescens* collected in north of the state of Durango México

Concept	Dry matter
In vitro gas production	31
Bypass protein	1
Microbial protein	6
Metabolizable protein	7
Potential dry matter digestibility	77
Potential digestible energy, Mcal/kg	3.6
Potential metabolizable energy, Mcal/kg	2.9
Potential dry matter intake, g/kg/day	83

CHAPTER 24

Nutrition of *Bernardia myricaefolia* (Scheele) Wats.

Introduction

Bernardia myricaefolia is commonly known Bernardia western or Bernardia. It belongs to the family of the Euphorbiaceae. Synonymy is *Bernardia incana* and *Tyria myricifolia.* The native plant is referred as a landscape shrub for south Texas and northeastern Mexico. It is relatively cold resistant, thornless, and well adapted to warm temperature and rain scarcity. Average height and width ranges between 1.0 and 2.0 m. This plant adaptable to any particular soil. The range of *B. myricaefolia* prolongs from the north to Central Texas and south to Tamaulipas, Nuevo Leon, and Coahuila, Mexico. Because the high nitrogen containing, leaves are quickly consumed by browsers. When the native brush in the Thornscrub is not subjected to browsing pressure, some edible species such as *Bernardia myricaefolia* happen in larger amounts than in comparable stands of shrublands from another place. The taxonomic characteristics are listed in Table 24.1.

Bernardia myricaefolia

Table 24.1. Taxonomic characteristics of *Bernardia myricaefolia*

Rank	Scientific Name and Common Name
Kingdom	Plantae – Plants
Subkingdom	Tracheobionta – Vascular plants
Superdivision	Spermatophyta – Seed plants
Division	Magnoliophyta – Flowering plants
Class	Magnoliopsida – Dicotyledons
Subclass	Rosidae
Order	Euphorbiales
Family	Euphorbiaceae – Spurge family
Genus	*Bernardia* Mill. – myrtlecroton
Species	*Bernardia myricaefolia* (Scheele) S. Watson – mouse's eye

Nutritional value

The nutritional profile of leaves of the shrub *Bernardia myricaefolia* is shown in Tables 24.2, 24.3 and 24.4. The plant has high ash content, crude protein and degradable protein. It has also high cellular content, but low neutral detergent fiber and its constituents (cellulose, hemicellulose and lignin). The leaves have not condensed tannins. The acetic acid is higher than the propionic acid and the butyric acid that is lowest. The volatile fatty acids (VFA) are a waste product of anaerobic microbial metabolism in ruminant providing a major source of metabolizable energy. The removal of these acids is also vital for maintaining rumen environment and for the continued growth of cellulolytic organisms. Main VFA in order of abundance are acetic, propionic, butyric, isobutyric, valeric and isovaleric acids. The VFA production is strongly influenced by diet and type of methanogenic population in the rumen, protozoa may contribute significantly in the balance. Other acids may appear as products, for example, lactic acid is important when starch is part of the diet, which is fermented to acetate, propionate and butyrate.

The predicted dry matter digestibility, digestible energy, metabolizable energy and dry matter intake of the leaves from B. myricaefolia are high enough for the adult range small ruminant needs. However, Na, P, Cu resulted marginal lower to fulfill the metabolic requirements of adult range ruminants. Ca, Mg, K, Fe, Mn and Zn were sufficient amounts for adult range ruminant requirements.

Table 24.2. Seasonal concentrations of the chemical composition (%, dry matter) and molar proportions of volatile fatty acids in leaves of Bernardia myricaefolia collected in northeastern Mexico

Concept	Seasons				Annual mean
	Winter	Spring	Summer	Fall	
Organic matter	84	85	85	87	85
Ash	16	15	15	13	15
Crude protein	17	17	17	17	17
Degradable protein	13	12	11	11	12
Undegraded protein	4	5	6	6	5
Neutral detergent fiber	28	29	31	30	29
Insoluble neutral detergent fiber	10	10	9	9	10
Cellular content	72	71	69	70	71
Acid detergent fiber	26	26	25	25	26
Cellulose	17	18	18	18	18
Hemicellulose	2	4	5	4	4
Lignin	7	7	7	6	7

Condensed tannins	0	0	0	1	0
Acetic acid	70	64	74	70	69
Propionic acid	17	18	16	16	17
Butyric acid	13	18	10	14	14

Table 24.3. Seasonal variation of predicted dry mater digestibility (%), digestible energy (Mcal/kg), metabolizable energy (Mcal/kg) and dry matter intake (g/kg BW/day) of the leaves of *Bernardia myricaefolia* collected in northeastern Mexico

Concept	Seasons				Annual
	Winter	Spring	Summer	Fall	mean
Dry matter digestibility	65	65	66	66	65
Digestible energy	3.0	3.0	3.1	3.1	3.0
Metabolizable energy	2.5	2.5	2.5	2.5	2.5
Dry matter intake	84	84	84	84	84

Table. 24.4. Seasonal changes of the macro (g/kg, dry matter) and microminerals (mg/kg, dry matter) in leaves of *Bernardia myricaefolia* collected in northeastern Mexico

Minerals	Seasons				Annual
	Winter	Spring	Summer	Fall	mean
Macrominerals					
Ca	43	42	40	42	42
K	7	6	8	7	7
Mg	3	3	2	3	3
Na	<1	<1	<1	<1	<1
P	1	1	1	1	1
Microminerals					
Cu	2	5	2	4	3
Fe	66	75	70	72	71
Mn	27	37	37	29	33
Zn	29	38	36	36	35

CHAPTER 25

Nutrition of *Sideroxylon celastrinum* (Kunth) T.D. Penn.

Introduction

Sideroxylon celastrinum is a species that belongs to Sapotaceae family that is native to Texas. Common names are Saffron Plum and Coma. It is a spiny shrubby or small tree that ranges in height from 2 to 9 m. The green leaves are alternative or fascicled at the nodes. The fruits and seeds are consumed by several species of birds, and mammals such as raccoons and coyotes, while the leaves are browsed for white-tailed deer (*Odocoileus virginianus*) goats and sheep. Cattle eat the leaves and benefits from the protective cover that provides their foliage. The tree provides a protective cover for many species of wildlife. The taxonomic characteristics of *Sideroxylon celastrinum* are listed in Table 25.1.

Sideroxylon celastrinum

Table 20.1. Taxonomic characteristics of *Sideroxylon celastrinum*

Rank	Scientific Name and Common Name
Kingdom	Plantae – Plants
Subkingdom	Viridiplantae
Superdivision	Embryophyta
Division	Spermatophytina – spermatophytes, seed plants, phanérogames
Class	Magnoliopsida – Dicotyledons
Subclass	Asteranae
Order	Ericales
Family	Sapotaceae – sapodillas, sapotes
Genus	*Sideroxylon L. – bumelia, bully*
Species	*Sideroxylon celastrinum* (Kunth) T.D. Penn. – bumelia, saffron plum

Nutritional value

The nutritional profiles of leaves from *Sideroxylon celastrinum* are listed in Tables 25.2, 25.3 and 25.4. The plant has high organic matter, crude protein. The neutral detergent fiber, mostly composed by cellulose, is double as much as the acid detergent fiber. It has high lignin content. Lignin governs the firmness, forte and resistance of the construction of plants. The digestion of a feedstuff is regulated by the content of lignin (the more the lignin content, the minor the feedstuff. Decreasing the lignin amount of fiber and foliage centrals to significantly diminished costs of making fiber and enhanced digestion of foliage and forage. Nevertheless, the benefits of diminish lignin are counterpoise by the difficulty of plants with diminish lignin that are more easily confronted by microorganisms. Certainly, augmenting lignin quantity has been stimulated as a protection against pests. The monomeric structure of lignin stimuluses the stuffs of the plant physical structure. There are two chief kinds of lignin, quaiacyl and guaiacyl-syringyl. Guaiacyl lignin is typical of softwoods that are hardy affected by chemical or biological hydrolysis. Guaiacyl-syingyl typical of hardwoods. Lignin content augments as plant mature. Feedstuffs high in lignin are poorly digested and as well as of low value. Plant materials such as grasses, alfalfa hay or corn with low lignin or lignin with augmented guaiacyl-syringyl quantity (easily processed) may offer a great financial benefit in livestock production, providing that the molecular changes do not effect in vulnerability to destructive microorganisms.

The condensed tannins were low and the amount of gross energy value is comparable to other commercial carbohydrates. The predicted dry matter digestibility, digestible energy, metabolizable energy and dry matter intake were in high quantities for the metabolic requirements of range small ruminants. The amounts of P, Na and Cu were lower, but Ca, Mg, K, Fe, Mn and Zn were sufficient for the metabolic needs of small rage ruminants consuming leaves of *Sideroxylon celastrinum* during all seasons of the year.

Table 25.2. Seasonal changes of the chemical composition (%, dry matter) of leaves from *Sideroxylon celastrinum* collected in northeastern Mexico

Concept	Seasons				Annual mean
	Winter	Spring	Summer	Fall	
Organic matter	93	91	91	90	91
Ash	7	9	9	10	9
Crude protein	18	18	15	17	17
Degradable protein	12	11	9	9	10
Undegraded protein	6	7	6	8	7
Neutral detergent fiber	65	62	65	59	34
Insoluble neutral detergent fiber	6	7	6	8	7
Acid detergent fiber	29	25	30	25	66
Cellulose	30	34	31	29	30
Hemicellulose	16	14	15	14	15
Lignin	13	11	15	11	15
Condensed tannins	4	2	2	3	3
Gross energy, Mcal/kg	5	5	5	5	5

Table 25.3. Seasonal variation of predicted dry mater digestibility (%), digestible energy (Mcal/kg), metabolizable energy (Mcal/kg) and dry matter intake (g/kg BW/day) of the leaves of *Sideroxylon celastrinum* collected in northeastern Mexico

Concept	Seasons				Annual
	Winter	Spring	Summer	Fall	mean
Dry matter digestibility	63	66	61	66	32
Digestible energy	2.9	3.1	2.9	3.1	1.6
Metabolizable energy	2.4	2.5	2.4	2.5	1.3
Dry matter intake	81	81	81	81	81

Table 25.4. Seasonal content of macro (g/kg, dry matter) and Microminerals (mg/kg, dry matter) in leaves of *Sideroxylon celastrinum* collected in northeastern Mexico

Concept	Seasons				Annual
	Winter	Spring	Summer	Fall	mean
Macrominerals					
Ca	23	32	31	34	30
K	14	11	18	13	14
Mg	6	5	5	4	5
Na	<1	<1	<1	<1	<1
P	1	1	1	1	1
Microminerals					
Cu	2	2	3	4	3
Fe	161	192	132	167	163
Mn	29	55	58	97	62
Zn	35	45	40	32	38

CHAPTER 26

Nutrition of *Castela texana* T. & G. Rose

Introduction

Castela texana (Goatbush) belongs to Simaroubaceae family and is a medium-sized plant with spine in branches and short linear leaves that are extremely unpleasant tasting. It offers good protection for wildlife but its unpleasant taste provides it slight forage and wildlife feed importance. The bottoms of the leaves are too much silvery, a distinctive that characterizes it from the shrub *Ziziphus obtusifolia*. Goatbush is very drought tolerant. A new quassinoid, 11-*O-trans-p*-coumaroyl amarolide was isolated from *Castela texana*, and the structure was clarified by analysis spectroscopic. The compound is the first coumaroyl quassinoid derivative to has been quarantined from wildlife flora. The recognized substances glaucarubolone, chaparrinone, amarolide, holacanthone, chaparrin and 15-*O*-β-d-glucopyranosyl glaucarubol were also identified. All identified substances were proved for their cytotoxicity and antiprotozoal actions. The taxonomic characteristics of *Castela texana* are listed in Table 26.1.

Castela texana

Table 26.1. Taxonomic characteristics of *Castella texana*

Rank	Scientific Name and Common Name
Kingdom	Plantae – Plants
Subkingdom	Tracheobionta – Vascular plants
Superdivision	Spermatophyta – Seed plants
Division	Magnoliophyta – Flowering plants
Class	Magnoliopsida – Dicotyledons
Subclass	Rosidae
Order	Sapindales
Family	Simaroubaceae – Quassia family
Genus	*Castela erecta* Turp. – goatbush
Species Subspecies	*Castela erecta* Turp. ssp. *texana* (Torr. & A. Gray) Cronquist – Texan *goatbush*

Nutritional value

The nutritional characteristics of *Castela texana* are listed in Tables 26.2, 26.3 and 26.4. It has high organic matter, medium-high crude protein. Low neutral detergent fiber, mostly composed by lignin. It has low condensed tannins and the gross energy is similar to other shrubs. Energy may be lost from an animal in the way of chemical energy (feces, urine, and gas) and in the way of heat. Digestible energy represents for only fecal loss that is frequently the major and most inconstant loss but it does not sufficiently distinguish foods. Metabolizable energy (ME) represents for energy missing via feces, urine and gas and is superior to digestible energy but it does not represent for the second most inconstant energetic loss (heat). Net energy (NE) represents for all losses and hypothetically is the most precise technique for distinguishing foods.

The predicted dry matter digestibility, digestible energy, metabolizable energy and dry matter intake were in sufficient amounts to fulfill the metabolic needs of adult range ruminates consuming the shrub. Minerals such as Na, P and Cu were lower for adult ruminant needs; however, Ca, K, Mg, FE, Mn and Zn were sufficient for adult ruminant requirements consuming leaves of Castela texana, during all seasons of the year.

Table 26.2. Seasonal changes of the chemical composition (%, dry matter) and gross energy (Mcal/kg, dry matter) in leaves of Castela texana collected in northeastern Mexico

Concept	Seasons				Mean
	Winter	Spring	Summer	Fall	mean
Organic matter	91	92	93	91	92
Ash	10	8	7	9	8
Crude protein	14	15	16	14	15
Degradable protein	11	12	11	9	11
Undegraded protein	3	3	5	5	4
Neutral detergent fiber	42	46	43	43	43
Insoluble neutral detergent fiber	11	12	12	12	12
Cellular content	58	54	57	57	57
Acid detergent fiber	31	32	32	33	32
Cellulose	6	9	10	8	9
Hemicellulose	11	14	11	10	12
Lignin	23	22	20	22	21
Condensed tannins	0	4	0	0	1
Gross energy	5	5	5	5	5

Table 26.3. Seasonal variation of predicted dry mater digestibility (%), digestible energy (Mcal/kg), metabolizable energy (Mcal/kg) and dry matter intake (g/kg BW/day) of the leaves of *Castela texana* collected in northeastern Mexico

Concept	Seasons				Annual
	Winter	Spring	Summer	Fall	mean
Dry matter digestibility	60	60	60	59	60
Digestible energy	2.8	2.8	2.8	2.8	2.8
Metabolizable energy	2.3	2.3	2.3	2.3	2.3
Dry matter intake	83	82	83	83	83

Table 26.4. Seasonal content of macro (g/kg, dry matter) and microminerals (mg/kg, dry matter) in leaves of *Castela texana* collected in northeastern Mexico

Concept	Seasons				Annual
	Winter	Spring	Summer	Fall	mean
Macrominerals					
Ca	33	29	22	30	29
K	8	11	10	7	8
Mg	4	5	6	5	5
Na	<1	<1	<1	<1	<1
P	1	1	1	1	1
Microminerals					
Cu	4	5	3	4	4
Fe	127	93	102	100	105
Mn	47	51	101	67	78
Zn	46	56	40	45	47

CHAPTER 27

Nutrition of *Celtis ehrenbergiana* (Klotzsch) Liebm.

Introduction

Celtis ehrenbergiana belongs to the family Ulmaceae; it is commonly named spiny hackberry or desert hackberry. It develops under dry areas like deserts, Thornscrub, and Grasslands. *Celtis ehrenbergiana* is the only species that has thorns. It is a shrub or small tree up to 3 m tall, with thorns on the branches. Leaves are small, less than 3 cm long and 2 cm wide. Flowers are situated in cymes of 3 to 5 flowers. The drupes are orange, and are consumed by wildlife and people. It is a valuable food for wildlife; birds, raccoons, white - tailed deer and wild rabbits consume the fruit. Domestic animals such as cattle, goats and sheep eat the leaves and branches. The taxonomic characteristics of *Celtis ehrenbergiana* are listed in Table 27.1.

Celtis ehrenbergiana

Table 27.1. Taxonomic characteristic of *Celtis ehrenbergiana*

Rank	Scientific Name and Common Name
Kingdom	Plantae – Plants
Subkingdom	Tracheobionta – Vascular plants
Superdivision	Spermatophyta – Seed plants
Division	Magnoliophyta – Flowering plants
Class	Magnoliopsida – Dicotyledons
Subclass	Hamamelididae
Order	Urticales
Family	Ulmaceae – Elm family
Genus	*Celtis* L. – hackberry
Species	*Celtis ehrenbergiana* (Klotzsch) Liebm. – spiny hackberry

Nutritional value

The chemical characteristics of the shrub are listed in Tables 27.2, 27.3 and 27.4. In general, the species has high ash and crude protein content; low organic matter and neutral detergent fiber and its components (cellulose, hemicellulose and lignin). It has not condensed tannins. The acetic acid is almost three times as higher as propionic acid and butyric acid was lowest. In general, the acetic and butyric acids are the major sources of energy, while propionic is reserved for gluconeogenesis. The acetic is the component most absorbed (90%) from the rumen, besides being the most important biolipogenic precursor. The volatile fatty acids (VFA) are absorbed in free form, and in their metabolism pass through the rumen wall into the portal blood, passing as anions neutralized in blood pH. The rumen epithelium considerably metabolizes VFA, resulting in higher quantities of butyrate and lower for acetate. Portal blood flow through liver, which takes most of propionate and butyrate, so all the acetate accounts for 90 % or more of the VFA that are in the peripheral circulation. The VFA give approximately 70% to the energy needs for ruminants, such as cattle and sheep, about 10% for humans, and around 20-30% for some other animals. The quantity of fiber in the diet unquestionably disturbs the quantity of VFA produced, and thus the contribution of VFA to the energy needs of the body could become noticeably greater as the nutritional fiber rises.

The predicted dry matter digestibility, digestible energy, metabolizable energy and dry matter intake of leaves from *Celtis ehrenbergiana* were in sufficient quantities to fulfill the metabolic requirements of adult range small ruminants. The Na, Cu and Zn were marginal below to satisfy small ruminant requirements. However, Ca, Mg, K, Mg, Mn and Fe were sufficient.

Table 27.2. Seasonal variations the chemical composition (5; dry matter) and molar proportions (Mol/100 Mol, dry matter) in leaves of *Celtis ehrenbergiana* collected in northeastern Mexico

Concept	Seasons				Annual
	Winter	Spring	Summer	Fall	mean
Organic matter	74	77	78	78	76
Ash	26	23	22	22	24
Crude protein	24	19	22	26	23
Degradable protein	20	15	16	21	17
Undegraded protein	4	4	6	5	5

Neutral detergent fiber	19	27	26	21	23
Insoluble neutral detergent fiber	6	6	6	6	6
Cellular content	81	73	74	79	77
Acid detergent fiber	14	16	21	16	17
Cellulose	8	11	13	10	11
Hemicellulose	5	10	5	5	6
Lignin	4	5	6	4	5
Condensed tannins	0	0	0	0	0
Acetic acid	70	70	77	74	73
Propionic acid	19	19	17	16	17
Butyric acid	11	11	6	10	10

Table 27.3. Seasonal variation of predicted dry mater digestibility (%), digestible energy (Mcal/kg), metabolizable energy (Mcal/kg) and dry matter intake (g/kg BW/day) of the leaves of *Celtis ehrenbergiana* collected in northeastern Mexico

Concept	Seasons				Annual
	Winter	Spring	Summer	Fall	mean
Dry matter digestibility	76	73	70	75	73
Digestible energy	3.5	3.4	3.3	3.5	3.4
Metabolizable energy	2.9	2.8	2.7	2.8	2.8
Dry matter intake	86	86	86	86	86

Table 27.4. Seasonal content of macro (g/kg, dry matter) and microminerals (mg/kg, dry matter) of leaves from *Celtis pallida* collected in northeastern Mexico

Concept	Seasons				Annual
	Winter	Spring	Summer	Fall	mean
Macrominerals					
Ca	40	43	49	49	45
K	15	13	19	16	16
Mn	7	6	7	7	7
Na	<1	<1	<1	<1	<1
P	2	1	2	2	2

Microminerals

Cu	4	7	3	4	5
Fe	77	67	54	99	74
Mn	33	28	48	51	32
Zn	20	30	22	22	19

CHAPTER 28

Nutrition of *Condalia hookeri* Hook.

Introduction

The species is an evergreen and spiny small tree or shrub, which develops in brushes and Thornscrub. It belongs to Rhamnaceae family. It propagates most plentifully in semiarid regions of south Texas and northeastern Mexico. The environment in which *Condalia hookeri* be greatly affects their nutritional properties. It is therefore very useful to know how and to what extent modify the physical and chemical properties of this plant, to manipulate and achieve higher quality forage production. Condalia species are found between 50 and 2,400 meters above sea level, mostly of them in xerophytic scrublands and tropical deciduous forest. The taxonomic characteristics are listed in Table 28.1.

Condalia hookeri

Table 28.1. Taxonomic characteristics of *Condalia hookeri*

Rank	Scientific Name and Common Name
Kingdom	Plantae – Plants
Subkingdom	Tracheobionta – Vascular plants
Superdivision	Spermatophyta – Seed plants
Division	Magnoliophyta – Flowering plants
Class	Magnoliopsida – Dicotyledons
Subclass	Rosidae
Order	Rhamnales
Family	Rhamnaceae – Buckthorn family
Genus	*Condalia* Cav. – snakewood
Species	*Condalia hookeri* M.C. Johnst. – Brazilian bluewood

Nutritional value

Tables 28.2, 28.3 and 28.4 show the nutritional profile of the plant *Condalia hookeri*. In general, it has high ash and crude protein content; low neutral detergent fiber and its constituents (cellulose, hemicellulose and lignin). Hemicellulose is higher than cellulose or lignin. It has low concentration of condensed tannins. The acetic acid is higher than propionic or butyric acids. About 50 to 60% of volatile fatty acids formed is acetic acid. It dominates on a high fiber ration and is a precursor for fat of milk. A little is utilized for fat and muscle tissues metabolism. Acetic acid was determined to be the main metabolite of the mammary gland of cows. In ruminants, the main product of carbohydrate digestion is acetic acid. This is taken and used for energy or for lipogenesis by various body tissues. Acetic acid must pass through acetyl CoA to be used as energy through oxidation in the citric acid cycle to yield 12 moles of ATP to be oxidized. Thus, there is a net gain of 10 ATP per mole of acetic acid absorbed. Acetic may pass through AGV synthesis by malonyl CoA carboxylation or alternatively through the citric acid cycle by condensing with oxaloacetate.

The predicted dry matter digestibility, digestible energy, metabolizable energy and dry matter intake were high enough for range small ruminant needs. The Na, P, Mn and Zn were marginal low for the metabolic requirements of adult range small ruminants. However, Ca, K, Mg, Cu and Fe were sufficient.

Table. 28.2. Seasonal concentrations of the chemical composition (%, dry matter) and molar proportions of volatile fatty acids (Mol/100 Mol, dry matter) in leaves of *Condalia hookeri* collected in northeastern Mexico

Concept	Seasons				Annual
	Winter	Spring	Summer	Fall	mean
Organic matter	80	80	79	79	80
Ash	20	20	21	21	20
Crude protein	16	19	15	15	16
Degradable protein	11	12	11	11	10
Undegraded protein	5	7	6	4	6
Neutral detergent fiber	45	39	44	42	42
Insoluble neutral detergent fiber	12	11	11	11	11
Cellular content	55	61	56	58	58
Acid detergent fiber	34	30	31	31	31
Cellulose	6	6	7	8	7
Hemicellulose	11	9	13	11	11
Lignin	4	6	5	6	5
Condensed tannins	0	1	0	0	0
Acetic acid	69	72	72	73	72
Propionic acid	23	21	22	20	20
Butyric acid	8	7	6	7	8

Table 28.3. Seasonal variation of predicted dry mater digestibility (%), digestible energy (Mcal/kg), metabolizable energy (Mcal/kg) and dry matter intake (g/kg BW/day) of the leaves of *Condalia hookeri* collected in northeastern Mexico

Concept	Seasons				Annual
	Winter	Spring	Summer	Fall	mean
Dry matter digestibility	58	62	60	60	61
Digestible energy	2.8	2.9	2.8	2.9	2.9
Metabolizable energy	2.3	2.4	2.3	2.3	2.3
Dry matter intake	82	83	83	83	83

Table 28.4. Seasonal content of macro (g/kg, dry matter) and microminerals (mg/kg, dry matter) in leaves of *Condalia hookeri* collected in northeastern Mexico

Concept	Seasons				Annual mean
	Winter	Spring	Summer	Fall	
Macrominerals					
Ca	9	10	11	10	10
K	16	18	16	17	17
Mg	12	14	13	13	13
Na	<1	<1	<1	<1	<1
P	1	1	1	1	1
Microminerals					
Cu	10	11	13	10	11
Fe	103	201	107	135	137
Mn	19	20	23	19	20
Zn	15	13	26	18	19

In Tables 28.5, 28.6 and 28.7 are listed the nutritional characteristics of leaves of *Condalia hookeri*, consumed by range ruminates and collected at the north region of Durango State, Mexico. It has good crude protein content, low neutral detergent fiber. The hemicellulose content is much higher than cellulose or lignin. It has very low values of condensed tannins and ether extract, the acetic acid is very low; however, is much higher than propionic or butyric, isobutyric, isovaleric valeric acids. Except of Na and Zn, other minerals were sufficient to satisfy the adult small ruminant metabolic requirements. The *in vitro* gas production was low along with the bypass protein, microbial protein and metabolizable protein. The potential dry matter digestibility was high as well as the potential digestible energy, potential metabolizable energy and potential dry matter intake values were sufficient for the needs of range ruminant consuming leaves of *Condalia hookeri*.

Table 28.5. Chemical composition (%) and *in vitro* volatile fatty acids (mM) of leaves from *Condalia hookeri* collected in the state of Durango México

Concept	Dry matter basis
Ash	9
Crude protein	14
Neutral detergent fiber	34
Acid detergent fiber	19
Lignin	10
Hemicellulose	14
Cellulose	9
Condensed tannins	5
Ether extract	2
Acetic	9
Propionic	1
Butyric	1
Isobutyric	0.1
Isovaleric	0.1
Valeric	0.1
Total	12
Acetic:Propionic	7

Table 28.6. Mineral composition of macro (g/kg, dry matter) and microminerals (mg/kg, dry matter) of leaves from *Condalia hookeri* collected in the state of Durango México

Concept	Minerals
Macrominerals	
Ca	41
K	4
Mg	4
Na	<1
P	4

Microminerals	
Cu	7
Fe	193
Mn	45
Zn	17

Table 28.7. *In vitro* gas production in 24 hours (GasP$_{24h}$/ mL 200 mg, dry matter), bypass protein (%, dry matter), microbial protein (%, dry matter) and metabolizable protein (%, dry matter) in leaves from *Condalia hookeri* collected in north of the state of Durango México

Concept	Dry matter
In vitro gas production	18
Bypass protein	1
Microbial protein	7
Metabolizable protein	8
Potential dry matter digestibility	70
Potential digestible energy, Mcal/kg	3.3
Potential metabolizable energy, Mcal/kg	2.7
Potential dry matter intake, g/kg/day	83

CHAPTER 29

Nutrition of *Cordia boisseri* A. DC.

Introduction

Cordia boisseri is a shrub or small tree of the Boraginaceae family. It grows from southern Texas, USA to northeastern Mexico. *Cordia boissieri* is an attractive shrub or tree with large, soft, dark leaves and trumpet-shaped white flowers. It is dry-tolerant abundant that it is a communal road establishing. White-tailed deer, sheep, goats and cattle like the fruit. In Mexico, leaves are utilized as teas for rheumatism and bronchial blocking. The leaves are ingested as fodder. The taxonomic characteristics of *Cordia boisseri* are listed in Table 29.1.

Cordia boisseri

Table 29.1. Taxonomic characteristics of *Cordia boisseri*

Rank	Scientific Name and Common Name
Kingdom	Plantae – Plants
Subkingdom	Tracheobionta – Vascular plants
Superdivision	Spermatophyta – Seed plants
Division	Magnoliophyta – Flowering plants
Class	Magnoliopsida – Dicotyledons
Subclass	Asteridae
Order	Lamiales
Family	Boraginaceae – Borage family
Genus	*Cordia* L. – cordia
Species	*Cordia boisseri* A. DC. – anacahuita

Nutritional value

Tables 29.2, 29.3 and 29.4 enlist the nutritional profile of *Cordia boissieri* collect in the Tamaulipan Thornscrub Vegetation in northeastern Mexico. In general, it has high ash and crude protein content; low neutral detergent fiber and its components (cellulose, hemicellulose and lignin). Cellulose content is almost three times as high as hemicellulose. It has not condensed tannins. The acetic acid is higher than propionic or butyric acids. Propionic acid is the primary glycogenic substrate. In ruminants, a large amount of this acid is obtained as a result of the breakdown of carbohydrates in the rumen. Most of this acid is removed from the portal blood by the liver. For the conversion of propionate to glucose, it requires that it first enter the Krebs cycle as succinyl CoA. In peripheral blood, there are small amounts of propionate that may precede insufficient capture of the liver or fatty acid oxidation odd number of carbon atoms. This propionate can be used to produce energy directly, with a net gain of 18 ATP. It is way is more efficient than the conversion of glucose.

The predicted dry matter intake, digestible energy, metabolizable energy and dry matter intake of leaves from *Cordia boisseri* were in sufficient amounts to fulfill metabolic requirements of adult range small ruminants. The Na, P, Mn and Zn were marginal lower to fulfill the metabolic requirements of adult ruminants. However, Ca, K, Mg, Cu and Fe were in sufficient amounts for adult ruminant needs.

Table 29.2. Seasonal variations of the chemical composition (%; dry matter) and molar proportions of volatile fatty acids (Mol/100 Mol, dry matter) in leaves of *Cordia boisseri* collected in northeastern Mexico

Concept	Seasons				Annual
	Winter	Spring	Summer	Fall	mean
Organic matter	82	84	86	87	86
Ash	18	16	14	13	16
Crude protein	19	14	12	14	15
Degradable protein	10	8	7	8	9
Undegraded protein	9	6	5	6	6
Neutral detergent fiber	36	36	37	38	37
Insoluble neutral detergent fiber	10	8	7	8	9
Cellular content	64	64	63	62	63
Acid detergent fiber	29	26	31	34	30
Cellulose	20	21	18	24	20
Hemicellulose	7	4	6	4	7
Lignin	5	10	12	9	9
Condensed tannins	0	0	0	0	0
Acetic acid	65	73	72	64	68
Propionic acid	28	23	23	32	27
Butyric acid	7	4	5	4	5

Table 29.3. Seasonal variation of predicted dry mater digestibility (%), digestible energy (Mcal/kg), metabolizable energy (Mcal/kg) and dry matter intake (g/ kg BW/day) of the leaves of *Cordia boisseri* collected in northeastern Mexico

Concept	Seasons				Annual
	Winter	Spring	Summer	Fall	mean
Dry matter digestibility	63	64	61	58	61
Digestible energy	3.0	3.0	2.9	2.8	2.9
Metabolizable energy	2.4	2.5	2.3	2.3	2.4
Dry matter intake	83	83	83	83	83

Table 29.4. Seasonal variations of macro (g/kg, dry matter) and microminerals (mg/kg, dry matter) in leaves of *Cordia boisseri* collected in northeastern Mexico

Concepto[1]	Seasons				Annual mean
	Winter	Spring	Summer	Fall	
Macrominerals					
Ca	17	12	12	14	14
K	10	16	11	12	13
Mg	9	8	8	7	8
Na	1	1	1	1	1
P	1	2	1	1	1
Microminerals					
Cu	10	11	14	11	11
Fe	284	192	303	256	259
Mn	21	16	20	19	20
Zn	15	35	21	23	24

In Tables 29.5, 29.6 and 29.7 are listed the nutritional characteristics of leaves of *Cordia boisseri*, consumed by range ruminates and collected at the north region of Durango State, Mexico. It has medium crude protein content, low neutral detergent fiber, cellulose values are higher than lignin or hemicellulose, very low values of condensed tannins and ether extract, the acetic acid is is much higher than propionic or butyric, isobutyric, isovaleric valeric acids. Except of Na and Cu, other minerals were sufficient for the needs of the adult small ruminant metabolic requirements. The *in vitro* gas production was low along with the bypass protein, microbial protein and metabolizable protein. The potential dry matter digestibility was also medium; however, the potential digestible energy, potential metabolizable energy and potential dry matter intake values were sufficient for the needs of range ruminant consuming leaves of *Cordia boisseri*.

Table 29.5. Chemical composition (%) and *in vitro* volatile fatty acids (mM) of leaves from *Cordia boisseri* collected in the state of Durango México

Concept	Dry matter basis
Ash	16
Crude protein	13
Neutral detergent fiber	41
Acid detergent fiber	32
Lignin	11
Hemicellulose	10
Cellulose	21
Condensed tannins	2
Ether extract	1
Acetic	15
Propionic	2
Butyric	2
Isobutyric	0.3
Isovaleric	0.4
Valeric	1
Total	21
Acetic:Propionic	7

Table 29.6. Mineral composition of macro (g/kg, dry matter) and microminerals (mg/kg, dry matter) of leaves from *Cordia boisseri* collected in the state of Durango México

Concept	Minerals
Macrominerals	
Ca	18
K	5
Mg	3
Na	1
P	3

Microminerals

Cu	4
Fe	285
Mn	47
Zn	32

Table 29.7. *In vitro* gas production in 24 hours ($GasP_{24h}$/ mL 200 mg, dry matter), bypass protein (%, dry matter), microbial protein (%, dry matter) and metabolizable protein (%, dry matter) in leaves from *Cordia boisseri* collected in north of the state of Durango México

Concept	Dry matter
In vitro gas production	29
Bypass protein	2
Microbial protein	6
Metabolizable protein	7
Potential dry matter digestibility	60
Potential digestible energy, Mcal/kg	2.8
Potential metabolizable energy, Mcal/kg	2.3
Potential dry matter intake, g/kg/day	83

CHAPTER 30

Nutrition of *Diospyros texana* Scheele.

Introduction

Diospyros texana belongs to the family Ebenaceae, in Mexico is known as chapote and in United State of America, it is called black persimmon. *Diospyros texana* is a species of persimmon that is native to Texas in the United States, and northeastern Mexico. The fruit is black. The berries are edible when they are black. Wildlife use Texas persimmon for food, shelter, and cover. Several mammals and birds consume the fruit. Spanish goats eat great quantities of the plant fodder, and white-tailed deer browse the foliage. Leaves of Texas persimmon averages 14.0% crude protein, 0.25% of P, 1.6% K, 2.5% Ca, 0.7% Mg, and 0.1% Na. The browse has medium food value for white-tailed deer. In thorn habitats, the plant and other browse plants form a high overlapping canopy that yields warm air, hiding, and discharge cover habitat for white-tailed deer. In Table 30.1 are listed the taxonomic characteristics of *Diospyros texana*.

Diospyros texana

Table 30.1. Taxonomic characteristics of *Diospyros texana*

Rank	Scientific Name and Common Name
Kingdom	Plantae – Plants
Subkingdom	Tracheobionta – Vascular plants
Superdivision	Spermatophyta – Seed plants
Division	Magnoliophyta – Flowering plants
Class	Magnoliopsida – Dicotyledons
Subclass	Dilleniidae
Order	Ebenales
Family	Ebenaceae – Ebony family
Genus	*Diospyros* L. – diospyros
Species	*Diospyros texana* Scheele – Texas persimmon

Nutritional value

In Tables 30.2, 30.3 and 30.4 are listed the nutritional characteristics of the species. In general, high organic matter medium crude protein, low undegraded protein and neutral detergent fiber and its components (cellulose, hemicellulose and lignin), low condensed tannins. Acetic acid is more than three times greater than propionic acid or butyric acid that was lowest. Upon passage through the rumen and omasum, butyric acid becomes β-hydroxybutyric acid. This can be used as an energy source by a number of tissues, especially skeletal muscle. This β-hydroxybutyrate is eventually converted to acetyl CoA and subsequently metabolized in the Krebs cycle. Net production of butyric acid per mole is 25 ATP. Butyrate ketone body is absorbed as energy source for the animal, especially during physiological crises such as parturition, exercise and pathological conditions such as toxemia of pregnancy. In general, acetic acid and butyric acid are the major source of energy (oxidation), while propionic is reserved for gluconeogenesis. Acetic acid is the most important lipogenic precursor.

The predicted dry matter digestibility, digestible energy, metabolizable energy and dry matter intake are in sufficient amounts to satisfy the metabolic demands of adult range small ruminants. The minerals Na, P and Zn were marginal lower to satisfy the metabolic requirements of adult small ruminants; however, Ca, K, Mg, Cu, Fe and Mn were in enough quantities to fulfill metabolic needs of adult small ruminates consuming foliage from *Diospyros texana,* in all seasons of the year.

Table 30.2. Seasonal variations of the chemical composition (%, dry matter) and molar proportions (Mol/100 Mol, dry matter) in leaves of *Diospyros texana* collected in northeastern Mexico

Concept	Seasons				Annual
	Winter	Spring	Summer	Fall	mean
Organic matter	88	88	88	87	88
Ash	12	12	12	13	12
Crude protein	12	12	14	14	13
Degradable protein	9	8	10	10	9
Undegraded protein	3	4	4	4	4
Neutral detergent fiber	30	31	34	37	33
Insoluble neutral detergent fiber	9	9	11	11	10
Cellular content	70	69	66	63	67
Acid detergent fiber	23	24	30	30	25
Cellulose	14	13	16	16	15
Hemicellulose	7	12	4	7	7
Lignin	9	11	12	10	11
Condensed tannins	2	2	2	2	2
Acetic acid	70	65	71	69	69
Propionic acid	18	19	19	21	19
Butyric acid	12	16	10	10	12

Table 30.3. Seasonal variation of predicted dry mater digestibility (%), digestible energy (Mcal/kg), metabolizable energy (Mcal/kg) and dry matter intake (g/kg BW/day) of the leaves of *Diospyros texana* collected in northeastern Mexico

Concept	Seasons				Annual
	Winter	Spring	Summer	Fall	mean
Dry matter digestibility	67	66	61	61	65
Digestible energy	3.1	3.1	2.9	2.9	3.1
Metabolizable energy	2.6	2.5	2.4	2.4	2.5
Dry matter intake	84	84	83	83	84

Table 30.4. Seasonal content of macro (g/kg, dry matter) and microminerals (mg/kg, dry matter) in leaves of *Diospyros texana* collected in northeastern Mexico

Concept	Seasons				Annual
	Winter	Spring	Summer	Fall	mean
Macrominerals					
Ca	31	20	21	29	20
K	12	15	11	13	10
Mg	3	3	3	4	2
Na	<1	<1	<1	<1	<1
P	1	1	1	1	1
Microminerals					
Cu	8	10	12	11	8
Fe	130	93	94	151	93
Mn	107	74	84	146	82
Zn	15	13	20	15	12

CHAPTER 31

Nutrition of *Flourensia cernua* DC.

Introduction

It is a species of flowering plant commonly known in English as tarbush and in Spanish is hojasé. The species has been evaluated as a potential complemental forage for sheep. Because it has high crude protein, it appears alike to alfalfa hay in nutritional profile. Nonetheless, it has chemical constituents that diminish its palatability to animals, making it unpleasant and "spicy". Moreover, the fruits and flowers are poisonous to domestic livestock. Livestock indeed evade it. However, the leaves may be consumed in restraint for their nutritional profile; however, rations composed only of tarbush may be lethal. The taxonomic characteristic of *Flourensia cernua* are listed in Table 31.1.

Flourensia cernua

Table 31.1. Taxonomic characteristics of *Flourensia cernua*

Rank	Scientific Name and Common Name
Kingdom	Plantae – Plants
Subkingdom	Tracheobionta – Vascular plants
Superdivision	Spermatophyta – Seed plants
Division	Magnoliophyta – Flowering plants
Class	Magnoliopsida – Dicotyledons
Subclass	Asteridae
Order	Asterales
Family	Asteraceae/Compositae – Aster family
Genus	*Flourensia* DC. – tarbush
Species	*Flourensia cernua* DC. – tarbush

Nutritional value

In Tables 31.2, 31.3 and 31.4 are listed the nutritional characteristics of leaves of S *Flourensia cernua*, consumed by range ruminates and collected in summer at the north region of Durango State, Mexico. It has medium high crude protein content, low neutral detergent fiber. The cellulose and hemicellulose content are similar and higher than lignin. It has very low values of condensed tannins and high ether extract content, the acetic acid is low; however, is much higher than propionic or butyric, isobutyric, isovaleric valeric acids. Except of Na, Cu and Zn, other minerals were sufficient to satisfy the adult small ruminant metabolic requirements. The *in vitro* gas production was high, low bypass protein, high microbial protein and metabolizable protein. The potential dry matter digestibility was high as well as the potential digestible energy, potential metabolizable energy and potential dry matter intake values were sufficient for the needs of range ruminant consuming leaves of *Flourensia cernua*.

Table 31.2 Chemical composition (%) and *in vitro* volatile fatty acids (mM) of leaves from *Flourensia cernua* collected in the state of Durango México

Concept	Dry matter basis
Ash	11
Crude protein	15
Neutral detergent fiber	32
Acid detergent fiber	20
Lignin	8
Hemicellulose	12
Cellulose	12
Condensed tannins	0
Ether extract	6
Acetic	13
Propionic	2
Butyric	1
Isobutyric	0.4
Isovaleric	0.3
Valeric	0.3
Total	17
Acetic:Propionic	7

Table 31.3. Mineral composition of macro (g/kg, dry matter) and microminerals (mg/kg, dry matter) of leaves from *Flourensia cernua* collected in the state of Durango México

Concept	Minerals
Macrominerals	
Ca	68
K	7
Mg	6
Na	1
P	2

Microminerals	
Cu	4
Fe	167
Mn	58
Zn	17

Table 31.4. *In vitro* gas production in 24 hours ($GasP_{24h}$/ mL 200 mg, dry matter), bypass protein (%, dry matter), microbial protein (%, dry matter) and metabolizable protein (%, dry matter) in leaves from *Flourensia cernua* collected in north of the state of Durango México

Concept	Dry matter
In vitro gas production	34
Bypass protein	1
Microbial protein	7
Metabolizable protein	8
Potential dry matter digestibility	70
Potential digestible energy, Mcal/kg	3.2
Potential metabolizable energy, Mcal/kg	2.7
Potential dry matter intake, g/kg/day	84

CHAPTER 32

Nutrition of *Forestiera angustifolia* Torr.

Introduction

Forestiera angustifolia belongs to the family Oleaceae. It is known as Texas Forestiera. It is native to northeastern Mexico and Texas, USA. It is a component of mixed shrubs and usually found in well-drained soils, rocky slopes completely sunny slopes and streams, poor lands. It is an excellent food for large mammals, good food for birds, small mammals and birds. White-tailed deer, sheep, cattle and goats browse the foliage, and many mammals, including raccoons, coyotes, foxes, rabbits, squirrels, rats and mice, consume the fruit. The plant also serves as a kind of canopy for some animals, to hedge against predators and is an important source of food for bees. It is used as an ornamental shrub in saline soils or limestone and the wind exposed areas. The taxonomic characteristics are listed in Table 32.1.

Forestiera angustifolia

Table 32.1. Taxonomic characteristics of *Forestiera angustifolia*

Rank	Scientific Name and Common Name
Kingdom	Plantae – Plants
Subkingdom	Tracheobionta – Vascular plants
Superdivision	Spermatophyta – Seed plants
Division	Magnoliophyta – Flowering plants
Class	Magnoliopsida – Dicotyledons
Subclass	Asteridae
Order	Scrophulariales
Family	Oleaceae – Olive family
Genus	*Forestiera* Poir. – swampprivet
Species	*Forestiera angustifolia* Torr. – Texas swampprivet

Nutritional value

In Table 32.2, 32.3 and 32.4 are listed the nutritional characteristic of *Forestiera angustifolia*. In general, it has high organic matter, crude protein and degradable protein; low neutral detergent fiber. The hemicellulose is higher than cellulose or lignin. It has not condensed tannins and the gross energy is similar as other native shrubs growing in northeastern Mexico. The predicted dry matter digestibility, digestible energy, metabolizable energy and dry matter intake are in levels to be sufficient for adult ruminant needs.

The P and Na were marginal insufficient to satisfy the metabolic requirements of adult ruminants. However, Ca, Mg, K, Cu, Fe, Mn and Zn were in sufficient concentrations to satisfy the adult ruminant requirements. Ca is the most abundant mineral element in the body with about 98% being located in the skeleton, where Ca, together with P, is used to provide strength to bones. The rest of the Ca is found either in extra cellular fluids of the body bound to plasma proteins or in ionized form. The major biological function of calcium is for bones. Bones contain 99% of the calcium in body. Calcium is also necessary for muscle contraction, nerve conduction and blood clotting. The main deficiency symptoms is called rickets. Milk is relatively high in Ca, and lactating goats need adequate levels of calcium for milk production.

Hypocalcemia can be caused while lactating. This condition is produced by a metabolic disorder that leads to a shortage of Ca in the blood because of Ca is being utilized for milk yield.

Table 32.2. Seasonal variations of the chemical composition (%, dry matter) and molar proportions (Mol/100 Mol, dry matter) in leaves of *Forestiera angustifolia* collected in northeastern Mexico

Concept	Seasons				Annual
	Winter	Spring	Summer	Fall	mean
Organic matter	91	93	91	91	91
Ash	9	7	9	9	9
Crude protein	17	21	19	12	17
Degradable protein	13	16	15	7	12
Undegraded protein	4	5	4	5	5
Neutral detergent fiber	41	52	40	34	42
Insoluble neutral detergent fiber	8	9	7	6	8
Cellular content	49	48	60	66	58
Acid detergent fiber	19	23	18	15	19
Cellulose	5	10	8	5	7
Hemicellulose	22	28	22	19	23
Lignin	13	13	10	10	12
Condensed tannins	0	0	0	0	0
Gross energy Mcal/kg	5	5	5	5	5

Table 32.3. Seasonal variation of predicted dry mater digestibility (%), digestible energy (Mcal/kg), metabolizable energy (Mcal/kg) and dry matter intake (g/ kg BW/day) of the leaves of *Forestiera angustifolia* collected in northeastern Mexico

Concept	Seasons				Annual
	Winter	Spring	Summer	Fall	mean
Dry matter digestibility	71	68	72	73	71
Digestible energy	3.3	3.2	3.3	3.4	3.3
Metabolizable energy	2.7	2.6	2.7	2.8	2.7
Dry matter intake	83	82	83	83	83

Table 32.4. Seasonal content of macro (g/kg, dry matter) and microminerals (mg/kg, dry matter) in leaves of *Forestiera angustifolia* collected in northeastern Mexico

Concept	Seasons				Annual mean
	Winter	Spring	Summer	Fall	
Macrominerals					
Ca	26	22	26	37	28
K	9	19	20	8	14
Mg	6	4	8	5	6
Na	<1	<1	<1	<1	<1
P	1	2	2	1	1
Microminerals					
Cu	8	23	11	10	13
Fe	666	119	135	129	262
Mn	126	100	149	142	129
Zn	71	87	77	89	81

CHAPTER 33

Nutrition of *Guaiacum angustifolium* Enbelm.

Introduction

Guaiacum angustifolium belongs to the family Zygophillaceae. It is commonly known as lignum vitae or tree of life. It is a slight constituent frequently found in mixed-brush areas throughout southern Texas, USA and northeastern Mexico. This species is an important browse shrub for white-tailed deer and range goats and sheep. The leaves have 16 to 18% crude protein. Its compact form and dense evergreen foliage makes it excellent as cover for wildlife. The taxonomic characteristics of *Guaiacum angustifolium* are listed in Table 33.1.

Guaiacum angustifolium

Table 33.1. Taxonomic characteristics of *Guaiacum angustifolium*

Rank	Scientific Name and Common Name
Kingdom	Plantae – Plants
Subkingdom	Tracheobionta – Vascular plants
Superdivision	Spermatophyta – Seed plants
Division	Magnoliophyta – Flowering plants
Class	Magnoliopsida – Dicotyledons
Subclass	Rosidae
Order	Sapindales
Family	Zygophillaceae – Creosote-bush family
Genus	*Guaiacum* L. – lignum-vitae
Species	*Guaiacum angustifolium* Engelm. – Texas lignum-vitae

Nutritional value

The nutritional profile of the leaves from *Guaiacum angustifolium* collected in northeastern Mexico is shown in Tables 33.2, 33.3 and 33.4. In general, it has high organic matter, crude protein and degradable protein. However, it has low neutral detergent fiber, lignin and hemicellulose are similar, but both are higher than cellulose. It does not have condensed tannins. The acetic acid is twice as much as higher than propionic acid and butyric acid was lowest. The predicted values of dry matter digestibility, digestible energy, metabolizable energy and dry matter intake were in sufficient to satisfy adult range small ruminants.

The concentrations of Na, P, Cu and Zn (except in summer) were in marginal insufficient to fulfill the metabolic requirements of adult small ruminants. However, Ca, K, Mg, Fe and Mn were sufficient. The Zn is located in all animal tissues and is necessary by the immune system and for regular growth of skin tissue. Zinc is in addition indispensable for reproduction males. Deficiency symptoms are dermatitis, loss of hair, lesions in skin, swollen feet, and reduced hair development.

Table 33.2. Seasonal variations of the chemical composition (%, dry matter) and molar proportions (Mol/100 Mol, dry matter) in leaves of *Guaiacum angustifolium* collected in northeastern Mexico

Concept	Seasons				Annual
	Winter	Spring	Summer	Fall	mean
Organic matter	86	87	87	82	84
Ash	14	13	13	18	16
Crude protein	20	18	17	15	17
Degradable protein	12	11	11	10	11
Undegraded protein	8	7	6	5	6
Neutral detergent fiber	39	39	39	36	38
In soluble neutral detergent fiber	9	10	12	10	10
Cellular content	61	61	61	64	62
Acid detergent fiber	23	27	34	28	28
Cellulose	9	12	5	8	8
Hemicellulose	16	14	14	13	13
Lignin	14	14	19	15	15
Condensed tannins	0	0	1	0	1
Acetic acid	61	59	71	65	64
Propionic acid	37	34	25	32	32
Butyric acid	2	7	4	3	4

Table 33.3. Seasonal variation of predicted dry mater digestibility (%), digestible energy (Mcal/kg), metabolizable energy (Mcal/kg) and dry matter intake (g/kg BW/day) of the leaves of *Guaiacum angustifolium* collected in northeastern Mexico

Concept	Seasons				Annual
	Winter	Spring	Summer	Fall	mean
Dry matter digestibility	68	64	58	63	63
Digestible energy	3.2	3.0	2.8	3.0	3.0
Metabolizable energy	2.6	2.5	2.3	2.4	2.4
Dry matter intake	83	83	83	83	83

Table 33.4. Seasonal content of macro (g/kg, dry matter) and microminerals (mg/kg, dry matter) in leaves of *Guaiacum angustifolium* collected in northeastern Mexico

Concept	Seasons				Annual mean
	Winter	Spring	Summer	Fall	
Macrominerals					
Ca	12	12	12	9	11
K	9	10	13	10	11
Mg	6	4	8	5	6
Na	<1	<1	<1	<1	<1
P	1	1	1	1	1
Microminerals					
Cu	5	5	4	4	5
Fe	96	98	208	223	156
Mn	126	100	149	142	129
Zn	19	22	44	20	26

CHAPTER 34

Nutrition of *Gymnosperma glutinosum* (Spreng) Less.

Introduction

Gymnosperma glutinosum belongs to the family Asteraceae (Table 34.1). In Mexico is recognized as Tatalencho. In the United States of America, it is identified as Maricopa. The only identified species is that is native to three countries: southwestern of the United States (Arizona, New Mexico, and Texas), Guatemala and Mexico. *Gymnosperma glutinosum* is an essential, and an active herbal medicine, which is extensive utilized in Mexico for the control of diarrhea. Two bioactive compounds (-)-17-hydroxy-neo-clerod-3-en-15-oic acid (1) and 5,7-dihydroxy-3,6,8,2',4',5'-hexamethoxyflavone (2) were isolated in the majority of the bacterial strains. The hexane extract showed antifungal activity against all tested fungi. It has been also reported that *Gymnosperma glutinosum* contains essential oils, flavonoids and diterpenes, and recent studies have validated the ethnobotanical use of this plant in Mexico. In addition, it has been showed anti-tumor activity of the hexane extract of G. *glutinosum* leaves against L5178Y-R lymphoma cells.

Gymnosperma glutinosum

Table 34.1. Taxonomic characteristics of *Gymnosperma glutinosum*

Rank	Scientific Name and Common Name
Kingdom	<u>Plantae</u> – Plants
Subkingdom	<u>Tracheobionta</u> – Vascular plants
Superdivision	<u>Spermatophyta</u> – Seed plants
Division	<u>Magnoliophyta</u> – Flowering plants
Class	<u>Magnoliopsida</u> – Dicotyledons
Subclass	<u>Asteridae</u>
Order	Asterales
Family	<u>Asteraceae/Compositae</u> – Aster family
Genus	*Gymnosperma* Less. – gymnosperma
Species	*Gymnosperma glutinosum* (Spreng.) Less. – gumhead

Nutritional value

In Tables 34.2, 34.3 and 34.4. In general, the plant has high organic matter, medium crude protein and low neutral detergent fiber and its components (cellulose, hemicellulose and lignin). The condensed tannins content are in beneficial levels to promote ruminal microbial growth in animals consuming then plant. The acetic acid is more than three times greater than propionic acid and butyric acid was lowest. The predicted dry matter digestibility, digestible energy, metabolizable energy and dry matter intake are in amount to fulfill adult ruminant needs. The P, Na and Cu were in marginal insufficient amounts to satisfy the metabolic requirements of adult ruminants consuming the plant during all seasons (except spring) of the year. However, minerals such as Ca, Mg, K, Fe, Mn and Zn were in sufficient amounts to satisfy the metabolic requirements of adult small range ruminants consuming foliage from *Gymnosperma glutinosum* during all year round.

Almost 70% of Mg is found in bone tissue. Magnesium is essential for the skeletal growth, nervous and functions of the muscular system and for activation of enzymes. It is also strictly related with metabolism of Ca and P. In ruminants, the main deficiency symptom is grass tetany, regularly understood as fast-growing, lush, cool season grasses. Low-blood Mg levels are present in affected animals, display a loss of hunger, are nervous, stumble, have spasms and some pass away are the deficiency symptoms of animals consuming low amounts of Mg.

Table. 34.2. Seasonal variation of the chemical composition (%, dry matter) and molar proportions (Mol/100 Mol, dry matter) in leaves of *Gymnosperma glutinosum* collected in northeastern Mexico

Concept	Seasons				Annual
	Winter	Spring	Summer	Fall	Mean
Organic matter	91	91	93	91	91
Ash	9	9	7	9	9
Crude protein	19	14	12	14	15
Degradable protein	17	12	10	12	12
Undegraded protein	5	2	2	2	3
Neutral detergent fiber	26	24	32	29	25
Insoluble neutral detergent fiber	8	6	11	9	8
Cellular content	74	76	68	71	75
Acid detergent fiber	20	16	29	25	21
Cellulose	7	8	9	9	9
Hemicellulose	6	8	4	4	4
Lignin	8	8	10	11	9
Condensed tannins	4	4	5	5	5

Acetic acid	78	68	68	80	74
Propionic acid	20	28	29	19	24
Butyric acid	2	5	3	5	3

Table 34.3. Seasonal variation of predicted dry mater digestibility (%), digestible energy (Mcal/kg), metabolizable energy (Mcal/kg) and dry matter intake (g/kg BW/day) of the leaves of *Gymnosperma glutinosum* collected in northeastern Mexico

Concept	Seasons				Annual
	Winter	Spring	Summer	Fall	mean
Dry matter digestibility	69	73	63	65	69
Digestible energy	3.2	3.4	3.0	3.1	3.2
Metabolizable energy	2.6	2.8	2.4	2.5	2.6
Dry matter intake	84	84	84	84	84

Table 34.4. Seasonal content of macro (g/kg, dry matter) and microminerals (mg/kg, dry matter) in leaves of *Gymnosperma glutinosum* collected in northeastern Mexico

Concepto[1]	Seasons				Annual
	Winter	Spring	Summer	Fall	mean
Macrominerals					
Ca	13	9	10	12	11
K	18	16	23	22	20
Mg	2	2	4	2	3
Na	<1	<1	<1	<1	<1
P	1	1	1	1	1
Microminerals					
Cu	6	9	6	6	7
Fe	79	71	67	79	74
Mn	33	30	56	44	45
Zn	150	155	163	175	161

CHAPTER 35

Nutrition of *Helietta parvifolia* Gray (Benth)

Introduction

Helietta parvifolia belongs to the family Rutaceae and is commonly known as barreta. A shrub or small tree that can reach up to 8 m tall, with a trunk up to 15 cm in diameter. Its branches are erect and form a small irregular crown. All parts are strongly aromatic. It is a forest product (rural posts, fences, pens, wood, shelves, etc.) and to a lesser scale livestock for production. However, range ruminants avidly consume dry fallen leaves. It is a tree of sunshine, high temperature resistant, thrives in semi-arid climates with shallow soils and limestone with loam - loamy, well-drained texture, with a pH of 7-8 and a depth of 60 cm, with rainfall of 450 to 900 mm annually. It is resistant to frost. The taxonomic characteristics are listed in Table 35.1.

Helietta parvifolia

Table 35.1. Taxonomic characteristics of *Helietta parvifolia*

Rank	Scientific Name and Common Name
Kingdom	Plantae – Plants
Subkingdom	Tracheobionta – Vascular plants
Superdivision	Spermatophyta – Seed plants
Division	Magnoliophyta – Flowering plants
Class	Magnoliopsida – Dicotyledons
Subclass	Rosidae
Order	Sapindales
Family	Rutaceae – Rue family
Genus	*Helietta* Tul. – helietta
Species	*Helietta parvifolia* (A. Gray ex Hemsl.) Benth. – barreta

Nutritional value

In Tables 35.2, 35.3 and 35.4 are listed the nutritional characteristics of the shrub *Helietta parvifolia* collected in northeastern Mexico. It has high organic matter content, medium crude protein, but with high level of degradable protein. It has very low neutral detergent fiber content, which is mainly composed by cellulose instead hemicellulose or lignin. It does not has condensed tannins. The acetic acid is about three times higher than the propionic acid or butyric acid. The predicted dry matter digestibility, digestible energy, metabolizable energy and dry matter intake of leaves from *Helietta parvifolia* are in sufficient levels to satisfy the metabolic requirements of adult range small ruminants. It appears that levels of Na, P and Cu are marginal insufficient to fulfill the needs of range ruminants consuming the leaves of *Helietta parvifolia*. However, the concentrations of Ca, K, Mg, Fe, Mn and Zn are in sufficient amounts to satisfy the metabolic requirements of adult ruminants.

Table 35.2. Seasonal variation of the chemical composition (%, dry matter) and molar proportions (Mol/100 Mol, dry matter) in leaves of *Helieta parvifolia* collected in northeastern Mexico

Concept	Seasons				Annual
	Winter	Spring	Summer	Fall	mean
Organic matter	87	88	87	89	88
Ash	13	12	13	11	12
Crude protein	13	12	14	12	13
Degradable protein	11	10	12	9	11
Undegraded protein	2	2	2	3	2
Neutral detergent fiber	18	19	21	19	19
Insoluble neutral detergent fiber	7	7	8	7	7
Cellular content	82	81	79	81	81
Acid detergent fiber	18	17	20	19	18
Cellulose	13	14	15	15	14
Hemicellulose	2	2	2	2	2
Lignin	4	3	5	4	4
Condensed tannins	0	0	0	0	0
Acetic acid	70	68	75	70	71
Propionic acid	16	19	15	17	17
Butyric acid	14	13	10	13	12

Table 35.3. Seasonal variation of predicted dry mater digestibility (%), digestible energy (Mcal/kg), metabolizable energy (Mcal/kg) and dry matter intake (g/kg BW/day) of the leaves of *Helieta parvifolia* collected in northeastern Mexico

Concept	Seasons				Annual
	Winter	Spring	Summer	Fall	mean
Dry matter digestibility	71	71	69	70	71
Digestible energy	3.3	3.3	3.2	3.3	3.3
Metabolizable energy	2.7	2.7	2.7	2.7	2.7
Dry matter intake	85	85	85	85	85

Table 35.4. Seasonal content of macro (g/kg, dry matter) and microminerals (mg/kg, dry matter) in leaves of *Helieta parvifolia* collected in northeastern Mexico

Concept	Seasons				Annual mean
	Winter	Spring	Summer	Fall	
Macrominerals					
Ca	23	32	34	29	29
K	10	8	9	9	9
Mg	6	7	7	7	7
Na	<1	<1	<1	<1	<1
P	1	1	1	1	1
Microminerals					
Cu	4	4	5	4	4
Fe	60	54	49	63	52
Mn	54	34	45	51	46
Zn	40	39	52	50	43

CHAPTER 36

Nutrition of *Karwinskia humboldtiana* (Schult.) Zucc.

Introduction

Karwinskia humboldtiana is commonly known as coyotillo or Humboldt coyotillo. It is a shrub or small tree of the family Rhamnaceae. It is native to Texas, USA and Mexico. The leaves and seeds of the species have the quinones eleutherin and 7-methoxyeleutherin and the fruits have chrysophanol and β-amyrin, which are poisonous to livestock. The leaves and seeds are toxic to livestock that do not graze it only in extreme dry conditions. However, some wildlife, like coyotes, like to eat the fruit. The taxonomic characteristics of *Karwinskia humboldtiana* are listed in Table 36.1

Karwinskia humboldtiana

Table 36.1. Taxonomic characteristics of *Karwinskia humboldtiana*

Rank	Scientific Name and Common Name
Kingdom	Plantae – Plants
Subkingdom	Tracheobionta – Vascular plants
Superdivision	Spermatophyta – Seed plants
Division	Magnoliophyta – Flowering plants
Class	Magnoliopsida – Dicotyledons
Subclass	Rosidae
Order	Rhamnales
Family	Rhamnaceae – Buckthorn family
Genus	*Karwinskia* Zucc. – karwinskia
Species	*Karwinskia humboldtiana* (Schult.) Zucc. – coyotillo

Nutritional value

The nutrition al profile of the plant species is shown in Tables 36.2, 36.3 and 36.4. It appears that the leaves have high organic matter, crude protein content and degradable protein. It has low neutral detergent fiber. The hemicellulose content is higher than cellulose or lignin. It has very low condensed tannins content. The gross energy is similar to other shrubs leaves. The predicted dry matter digestibility, digestible energy, metabolizable energy and dry matter intake of leaves from *Karwinskia humboldtiana* collected in northeastern Mexico are in levels that can be sufficient for the metabolic requirements of adult range small ruminants.

Minerals such as Ca, K, Mg, Fe, Mn, and Zn were in sufficient amounts to fulfill the metabolic requirements of adult ruminants. However, Cu, Na and P were in marginal low quantities for the needs of range small ruminants. The P has the more known biological functions than any other mineral element. About 80% of P is localized in the bone tissue, in which it is existing along with Ca. It is also localized in the tissues in which it is associated with energy conversion. It is an important constituent of buffer systems in the blood and other fluids of the body. The rumen microbes commands nutritional needs for ruminants with P being essential for the digestibility of the fiber and proteins. The main symptoms of P deficiency are diminished growth, lethargy, unkempt presence, reduced fertility, depraved hunger, consuming wood, rocks and bones) and diminished P in serum. The P is the greatest frequently faced mineral deficiency and the maximum costly macromineral.

Table 36.2. Seasonal variation of the chemical composition (%, dry matter) in leaves of *Karwinskia humboldtiana* collected in northeastern Mexico

Concept	Seasons				Annual
	Winter	Spring	Summer	Fall	mean
Organic matter	86	87	87	85	86
Ash	14	13	13	15	14
Crude protein	18	18	21	26	21
Degradable protein	17	19	12	12	15
Undegraded protein	4	7	6	6	6
Neutral detergent fiber	35	40	37	35	37
Insoluble neutral detergent fiber	7	9	8	8	8
Cellular content	65	60	63	65	63
Acid detergent fiber	19	25	21	21	21
Cellulose	10	8	9	13	10
Hemicellulose	16	14	17	14	15
Lignin	10	17	10	8	11
Condensed tannins	3	1	3	2	2
Gross energy, Mcal/kg	5	5	5	5	5

Table 36.3. Seasonal variation of predicted dry mater digestibility (%), digestible energy (Mcal/kg), metabolizable energy (Mcal/kg) and dry matter intake (g/kg BW/day) of the leaves of *Karwinskia humboldtiana* collected in northeastern Mexico

Concept	Seasons				Annual
	Winter	Spring	Summer	Fall	mean
Dry matter digestibility	71	66	70	70	70
Digestible energy	3.3	3.1	3.3	3.3	3.3
Metabolizable energy	2.7	2.5	2.7	2.7	2.7
Dry matter intake	83	83	83	83	83

Table 36.4. Seasonal content of micro (g/kg, dry matter) and microminerals (mg/kg, dry matter) in leaves of *Karwinskia humboldtiana* collected in northeastern Mexico

Concept	Seasons				Annual mean
	Winter	Spring	Summer	Fall	
Macrominerals					
Ca	59	43	48	60	53
K	11	27	13	18	17
Mg	2	2	2	2	2
Na	<1	<1	<1	<1	<1
P	1	2	2	1	2
Microminerals					
Cu	5	7	8	6	7
Fe	167	127	147	229	243
Mn	76	80	87	70	78
Zn	44	44	30	30	37

CHAPTER 37

Nutrition of *Larrea tridentata* DC.

Introduction

Larrea tridentata is a plant that belongs to the family Zygophyllaceae. It is known as creosotebush and as gobernadora in Mexico, Spanish for governess, due to its aptitude to safe more water by preventing the growth of adjacent plants. Young creosotebush is much more vulnerable to dry stress than established plants. Germination is more vigorous during rainy times. Mature plants, nonetheless, may accept extreme dry stress. Cell division may happen during these times of water tension, and it is common for new cells to rapidly captivate water after rainfall. The quick uptake origins twigs to propagate some centimeters at the final of a rainy season. It has a large number of chemical compounds in their leaves, seemingly as an anti-herbivore strategy. The taxonomic characteristics of *Larrea tridentata* are listed in Table 37.1.

Larrea tridentata

Table 37.1. Taxonomic characteristics of *Larrea tridentata*

Rank	Scientific name and common nale
Kingdom	Plantae - Plants
Subkingdom	Tracheobionta – Vascular plants
Superdivision	Spermatophyta – Seed plants
Division	Magnoliophyta – Flowering plants
Class	Magnoliopsida – Dicotyledons
Subclass	Rosidae
Order	Sapindales
Family	Zygophyllaceae – Creosote-bush family
Genus	*Larrea* Cav. – creosote bush
Species	*Larrea tridentata* (DC.) Coville – creosote bush

Nutritional value

In Tables 37.2, 37.3 and 37.4 are listed the nutritional characteristic of *Larrea tridentata* collected in northeastern Mexico. In general, the plant has high organic matter, crude protein, degradable protein, low neutral detergent fiber; lignin is higher than cellulose or hemicellulose. It has very low content of condensed tannins and the gross energy is similar than other native shrubs.

The concentrations of Ca, K, Mg, Fe, Mn and Zn are enough to satisfy the metabolic requirements of adult range small ruminants; however, Na, P and Cu were lower. Copper is important in creation of red blood cells, hair coloration, connective tissue and as enzyme activator. It is also significant in regular protected system function and nerve transmission. Deficiency symptoms are anemia, lightened looking and coarse hair covering, diarrhea and mass loss. Young ruminants may have advanced incoordination and paralysis, particularly in the back legs. High dietary Mo can diminish absorption of Cu and promote a deficiency symptom. Hair and wool sheep are delicate to Cu toxicity; however, goats require Cu quantities comparable to cattle. Angora goats can be more delicate to Cu toxicity than other types of goats. There exist variances in Cu requirements for some kind of sheep. This may in addition be true for meat type goats. The predicted dry matter digestibility, digestible energy and metabolizable energy and dry matter intake values of fallen leaves of *Larrea tridentata* are sufficient for adult ruminant needs.

Table 37.2. Seasonal variation of the chemical composition (%, dry matter) in leaves of *Larrea tridentata* collected in northeastern Mexico

Concept	Seasons				Annual
	Winter	Spring	Summer	Fall	mean
Organic matter	87	86	84	88	87
Ash	13	14	16	12	13
Crude protein	17	18	17	18	17
Degradable protein	13	11	10	11	11
Undegraded protein	4	7	7	7	6
Neutral detergent fiber	24	27	19	18	22
Insoluble neutral detergent fiber	7	9	7	6	7
Cellular content	76	73	81	82	78
Acid detergent fiber	18	24	17	15	18
Cellulose	8	9	5	6	7
Hemicellulose	6	3	2	4	4
Lignin	10	13	9	8	10
Condensed tannins	1	1	0	2	1
Gross energy, Mcal/kg	5	5	5	5	5

Table 37.3. Seasonal variation of predicted dry mater digestibility (%), digestible energy (Mcal/kg), metabolizable energy (Mcal/kg) and dry matter intake (g/kg BW/day) of the leaves of *Larrea tridentata* collected in northeastern Mexico

Concept	Seasons				Annual
	Winter	Spring	Summer	Fall	mean
Dry matter digestibility	71	67	72	74	71
Digestible energy	3.3	3.1	3.4	3.4	3.3
Metabolizable energy	2.7	2.6	2.8	2.8	2.7
Dry matter intake	84	84	85	85	85

Table 37.4. Seasonal content of micro (g/kg, dry matter) and microminerals (mg/kg, dry matter) in leaves of *Larrea tridentata* collected in northeastern Mexico

Concept	Seasons				Annual mean
	Winter	Spring	Summer	Fall	
Macrominerals					
Ca	39	34	51	33	39
K	22	19	35	24	25
Mg	3	2	3	3	3
Na	1	1	1	1	1
P	1	1	1	1	1
Microminerals					
Cu	5	4	6	6	5
Fe	295	310	376	354	457
Mn	110	191	167	110	145
Zn	49	41	69	57	54

CHAPTER 38

Nutrition of *Leucophyllum frutescens* (Berl.) I.M. Johnst.

Introduction

Leucophyllum frutescens is an <u>evergreen</u> <u>shrub</u> belongs to the family <u>Scrophulariaceae</u>. It is native to the <u>Texas</u>, USA and northern <u>Mexico</u>. The plant species is named Texas Ranger or Cenizo. Because of the characteristics and nutritional value presented by the *Leucophyllum frutescens* is exploited as fodder for goats, sheep cattle and white-tailed deer. The fact that the annual rainfall in northeastern Nuevo Leon is 400 to 600 mm and the average annual temperature varies from 29 °C, in summer and 14 °C in winter, makes the northeastern region more appropriate for the exploitation of range livestock, being range goats that best adapted to these places. The taxonomic characteristics of *Leucophyllum, frutescens* are listed in Table 38.1.

Leucophyllum frutescens

Table 38.1. Taxonomic characteristics of *Leucophyllum frutescens*

Rank	Scientific Name and Common Name
Kingdom	Plantae – Plants
Subkingdom	Tracheobionta – Vascular plants
Superdivision	Spermatophyta – Seed plants
Division	Magnoliophyta – Flowering plants
Class	Magnoliopsida – Dicotyledons
Subclass	Asteridae
Order	Scrophulariales
Family	Scrophulariaceae – Figwort family
Genus	*Leucophyllum* Bonpl. – barometerbush
Species	*Leucophyllum frutescens* (Berl.) I.M. Johnst. – Texas barometer bush

Nutritional value

In Tables 38.2, 38.3 and 38.4 is shown the nutritional profile of fallen leaves from *Leucophyllum frutescens* collected in northeastern Mexico. It has high organic matter, medium crude protein content, low neutral detergent fiber and very high lignin content. It does not have condensed tannins. The acetic acid is more than three times greater than propionic acid. The butyric acid was lowest. The predicted levels of dry matter digestibility, digestible energy, metabolizable energy and dry matter intake of fallen leaves of *Leucophyllum frutescens* are in good enough amounts to satisfy adult ruminant needs during all season of the year.

Concentrations of Na, P and Mn were marginal lower to fulfill the metabolic requirements of adult range small ruminants: however, Ca, K, Mg, Cu Zn and Fe were in sufficient amounts. The main function of Fe is as a component of hemoglobin, which is required for oxygen transport. It is a part of certain enzymes as a cofactor. The main symptom of deficiency of Fe is anemia. Anemia may be produced by blood injury due to some circumstances, including wound, internal parasites. The Fe is deposited in the spleen, bone marrow and liver. Milk has very low content of Fe; therefore, kids drinking for a long time only milk may suffer anemia. Soil contamination on forages can provide important amounts of dietary Fe.

Table 38.2. Seasonal variation of the chemical composition (%, dry matter) and molar proportions (Mol/100 Mol, dry matter) of leaves from *Leucophyllum frutescens* collected in northeastern Mexico

Concept	Seasons				Annual
	Winter	Spring	Summer	Fall	mean
Organic matter	94	91	89	92	92
Ash	6	9	11	8	8
Crude protein	10	14	13	12	12
Degradable protein	6	8	8	7	7
Undegraded protein	4	6	5	5	5
Neutral detergent fiber	45	44	44	42	44
Insoluble neutral detergent fiber	12	12	12	12	12
Cellular content	55	56	56	58	56
Acid detergent fiber	34	34	33	32	33
Cellulose	8	11	8	7	8
Hemicellulose	11	10	11	10	11
Lignin	26	22	24	24	24
Condensed tannins	0	1	0	0	0
Acetic acid	66	72	69	68	69
Propionic acid	25	20	23	24	23
Butyric acid	9	8	8	8	8

Table 38.3. Seasonal variation of predicted dry mater digestibility (%), digestible energy (Mcal/kg), metabolizable energy (Mcal/kg) and dry matter intake (g/kg BW/day) of the leaves of *Leucophyllum frutescens* collected in northeastern Mexico

Concept	Seasons				Annual
	Winter	Spring	Summer	Fall	mean
Dry matter digestibility	57	58	58	59	58
Digestible energy	2.7	2.7	2.8	2.8	2.8
Metabolizable energy	2.2	2.3	2.3	2.3	2.3
Dry matter intake	82	83	83	83	83

Table 38.4. Season al content of macro (g/kg, dry matter) and microminerals (mg/kg, dry matter) in leaves of *Leucophyllum frutescens* collected in northeastern Mexico

Concept	Seasons				Annual
	Winter	Spring	Summer	Fall	mean
Macrominerals					
Ca	10	5	8	8	8
K	21	15	15	9	15
Mg	10	5	6	4	6
Na	<1	<1	<1	<1	<1
P	1	1	1	1	1
Microminerals					
Cu	15	9	19	18	15
Fe	288	355	363	448	365
Mn	29	24	22	25	23
Zn	39	40	41	65	40

CHAPTER 39

Nutrition of *Opuntia engelmannii* Salm-Dyck ex Engelm.

Introduction

Opuntia engelmannii belongs to the family Cactaceae is known in Mexico as forage nopal and in USA Texas pricklypear. This species is used as fodder for livestock in the dry season and is also used as a food supplement. Given the characteristics of arid and semi-arid prevailing in some areas of the region, pricklypear may be considered as an alternative source of fodder. Xerophytes plants have the characteristic to adapt and resist drought, some of these characters are directly related to the increased efficiency of absorption of water through a very wide shallow root system. Water is stored through a specialized with the regulation of transpiration by a thick cuticle and protected stomata. Another important feature of physiological type is the remarkable speed with which species react to. The taxonomic characteristics of *Opuntia engelmannii* are listed in Table 39.1.

Opuntia engelmannii

Table 39.1. Taxonomic characteristics of *Opuntia engelmannii*

Rank	Scientific Name and Common Name
Kingdom	Plantae – Plants
Subkingdom	Tracheobionta – Vascular plants
Superdivision	Spermatophyta – Seed plants
Division	Magnoliophyta – Flowering plants
Class	Magnoliopsida – Dicotyledons
Subclass	Caryophyllidae
Order	Caryophyllales
Family	Cactaceae – Cactus family
Genus	*Opuntia* Mill. – pricklypear
Species	*Opuntia engelmannii* Salm-Dyck ex Engelm. – cactus apple

Nutritional value

With a characteristic content of 90% water (fresh matter), 4 to 6% crude protein (dry matter), 4% Ca (dry matter), 75% in vitro dry matter digestibility, and 72% digestible protein, pricklypear gives a greatly digestible source of energy, a plenty source of Ca for animals under lactation, and high water content to counterbalance the animal water need during dry seasons. However, due to the low crude protein content, it is required to complement animal diets based on unfertilized pricklypear with crude protein, vitamin and mineral complements. The vitamin C contents are medium compared to other forages. Meanwhile, the carotenoid contents are not exceptional compared to other forages, in dry seasons when all other herbaceous forages are mature, pricklypear sometimes are the unique source of carotenoids.

In Tables 39.2, 39.3 and 39.4 are listed the nutritional characteristics of cladodes of *Opuntia engelmannii* collected in northeastern Mexico. In general, it has high ash and low crude protein content. Very low neutral detergent fiber. Hemicellulose is higher than cellulose or lignin content. The acetic acid is higher than propionic acid or butyric acid. The predicted levels of dry matter digestibility, digestible energy, metabolizable energy and dry matter intake of cladodes from *Opuntia engelmannii* collected in northeastern Mexico were in sufficient amounts to satisfy adult range small cattle.

The minerals Na, P and Zn were marginal insufficient to satisfy the metabolic adult ruminant needs. However, Can K, Mg, Cu, Fe and Mn were satisfactory. The Mn is essential for bone tissue formation, reproduction and as enzyme cofactor. Deficiency symptoms are a hesitancy to walk, malformation of forelegs, late onset of estrus, poor fertility rate and low weight at birth. However, Mn deficiencies are uncommon.

Table. 39.2. Seasonal variations of the chemical composition (%, dry matter) and molar proportions (Mol/100 Mol, dry matter) in cladodes of *Opuntia engelmannii* collected in northeastern Mexico

Concept	Seasons				Annual
	Winter	Spring	Summer	Fall	mean
Organic matter	80	75	79	76	77
Ash	20	25	21	24	23
Crude protein	5	4	6	5	5
Degradable protein	3	3	4	3	3
Undegraded protein	2	1	2	2	2

Neutral detergent fiber	37	36	37	37	37
Insoluble neutral detergent fiber	6	6	6	6	6
Cellular content	63	64	63	63	63
Acid detergent fiber	13	15	14	13	14
Cellulose	12	13	13	13	13
Hemicellulose	24	22	24	24	23
Lignin	1	2	1	1	1
Condensed tannins	0	0	0	0	0
Acetic acid	69	70	77	74	72
Propionic acid	24	23	17	20	21
Butyric acid	7	7	6	7	7

Table 39.3. Seasonal variation of predicted dry mater digestibility (%), digestible energy (Mcal/kg), metabolizable energy (Mcal/kg) and dry matter intake (g/kg BW/day) of the leaves of *Opuntia engelmannii* collected in northeastern Mexico

Concept	Seasons				Annual
	Winter	Spring	Summer	Fall	mean
Dry matter digestibility	74	72	73	74	73
Digestible energy	3.4	3.3	3.4	3.4	3.4
Metabolizable energy	2.8	2.7	2.8	2.8	2.8
Dry matter intake	83	83	83	83	83

Table 39.4. Seasonal content of macro (k/kg, dry matter) and microminerals (mg/kg, dry matter) in cladodes of *Opuntia engelmannii* collected in northeastern Mexico

Concept	Seasons				Annual
	Winter	Spring	Summer	Fall	mean
Macrominerals					
Ca	23	27	27	26	25
K	13	17	22	15	17
Mg	15	14	18	12	15
Na	<1	<1	<1	<1	<1
P	<1	<1	<1	<1	<1

Microminerals

Cu	16	15	17	21	17
Fe	43	75	40	97	64
Mn	32	42	41	39	40
Zn	25	16	16	17	18

CHAPTER 40

Nutrition of *Quercus* spp

Introduction

Quercus spp (oak) is a tree or shrub of Fagaceae family. There exist approximately 600 existing species of oaks. The foliage of *Quercus* spp can be used as food because it has presence of foliage for most of the year and biomass during the dry season, and the content of secondary compounds are in moderate amounts presenting beneficial effects in animals. Oak leaves and twigs are often grazed by ruminants or harvested for use as livestock feed during feed shortages. Leaf materials may be used as a forage source for maintenance and to provision a few growth in goat kids so long as complemented with mineral elements. Small ruminant animals consume the oak leaves and nuts to meet their energy, protein and mineral requirements in most parts of world. In Table 40.1 are listed the taxonomic features of *Quercus* spp.

Quercus spp

Table 40.1. Taxonomic characteristics of *Quercus* spp

Rank	Scientific Name and Common Name
Kingdom	Plantae – Plants
Subkingdom	Tracheobionta – Vascular plants
Superdivision	Spermatophyta – Seed plants
Division	Magnoliophyta – Flowering plants
Class	Magnoliopsida – Dicotyledons
Subclass	Hamamelididae
Order	Fagales
Family	Fagaceae – Beech family
Genus	*Quercus* L. – oak

Nutritional value

It has been noted that the leaves of *Quercus* contain crude protein (CP) content to be used as a supplement for grazing animals on diets of poor quality based on cereal straws (Table 40.2). However, the nutritional composition varies across the different species an d seasons of the year.

Table 40.2. Chemical composition (%, dry matter) of some *Quercus* species

Concept	*Quercus persica*	*Quercus infectoria*	*Quercus libani*	*Quercus gambelii*
Crude protein	11.5	9.2	12.3	
Neutral detergent fiber	53.2	54.0	51.20	
Acid detergent fiber	31.7	30.0	33.1	28.9
Lignin	9.8	10.3	9.5	10.2
Total tannins	7.3	10.9	10.0	11.1
Hydrolyzed tannins	4.6	8.7	6.2	
Condensed tannins	1.4	1.5	1.2	
Effective degradability of dry matter	30.0	37.7	37.5	

In general, it is been mentioned that *in vivo* digestibility of goats feeding species of *Quercus,* collected in Europe and North America, appeared to be in a range of 47 to 49%, and could rise 4-9% if included in the diet a high quality forages such as alfalfa hay (*Medicago sativa*) or soybeans.

When it is supplied *Q. calliprinos* as a single component of the diet of pregnant adult goats are slightly deficient in CP, but it provides enough energy for maintenance. Poor digestibility of the foliage of *Quercus* is generally attributed to the high content of tannins found in all species. It has been observed that immature leaves of *Q. gambelii* contained 11.1% tannin and 8.7% mature while other authors mention that *Q. grisea* contained 7.9 % and 5.4%, respectively,

In addition, conducted studies in order to analyze the effect of soil type and geographic location on the content of macrominerales in leaves of *Q. robur* L. there was no individual variability between soil types, while the location significantly influenced. Therefore, it was concluded that the environment mainly influenced variability in macroelements content in the leaves.

It has been mentioned that cattle, sheep, goats, horses and pigs avidly consume many leaves and acorns of *Quercus* species. The parts of the plant and the season, when livestock consume it more easily, depends largely on the growth habits of each species of *Quercus*. In Europe, it has been reported that leaves and branches of *Q. flex* can be used as a protein supplement for goats. In southwestern sates of USA, where *Q. grisea*, *Q. dumosa* and *Q. gambelii* are common, leaves, sprouts and twigs are avidly browsed by goats and to a lesser extent sheep and cattle. Similar responses has been reported in southern Europe for *Q. flex* and *Q. pubescens*.

Meanwhile in Israel, leaves and young stems of *Q. calliprinos* are a major source of feed for cattle and goats. Reported levels of native *Quercus* on voluntary intake by goats in France, Greece and Israel are higher than with the common species in North America, at equivalent levels of CP content and digestibility. In India and Nepal the leaves of some species such as *Q. incana*, *Q. glauca* and *Q. semecarpifolia* are harvested for direct feeding in fresh or for storage for use in times of shortage of fodder. In such situations, the importance of acorns in the feed is limited than leaves. In addition, it has been mentioned that in some areas, fallen leaves on the ground under the trees are consumed, but this material is not particularly attractive, and is of poor quality and low digestibility.

The use of leaves as fodder has been reported to Nuevo Leon and San Luis Potosi, Mexico. It has been found that *Quercus* leaves are a used as a staple food for goats during most of the year. In Tlaxcala, the fresh leaves of *Q. crassipes* and Q. obtusata are used to feed donkeys.

It has been noted that when leaves of *Quercus* represented a large part of the diet of goats, its low digestibility can reduce voluntary intake. This, in turn, results in a diet deficient in metabolizable energy (ME), especially when consumed along with a native poor forage quality. However, when used as a supplement in 25% of the diet, Q. incana was able to increase digestibility and voluntary intake of wheat straw. *Quercus* leaves are used more effectively when included at low levels to supplement foods of poor quality.

In forests where the predominant plant species is *Q. wislizenii* and in seasons that there is no availability of grasses, 84% of food intake of sheep, is represented by the biomass of Quercus. While it is not common practice, the use of sawdust of Quercus, has been studied as a component of fiber in high-energy diets for finalization cattle. The sawdust could properly be used at levels up to 15% as fed in a ration based on corn bran and soybean meal, without adverse effects on growth rates or carcass characteristics. It has been reported that the acorns of *Q. coccifera* are a profitable source of energy for small ruminants and can replace 50% barley concentrate a basal diet of goats. Conversely, it was suggested that replacing barley by acorns should be at a maximum level of 25%.

In a study carried out with the aim to evaluate the chemical composition and degradability of foliage from *Quercus resinosa*, the biomass yield of *Quercus rugosa* and the effects of inclusion in diets of sheep, intake and palatability. Sample collections were conducted during two years at three sites: site 1 in the town of Tequila, Jalisco, Mexico; site 2 was located in the town of Teul Gonzalez Ortega, Zacatecas, Mexico; and the site 3 in the town of San Cristobal de la Barranca, Jalisco, Mexico. The values of organic matter, crude protein, non-fiber carbohydrates, ether extract, metabolizable energy, net energy for lactancy, short chain fatty acids condensed tannins *in vitro* organic matter digestibility and *in situ* degradability parameters of *Q. resinosa* leaves were higher in young leaves than in mature. However, neutral detergent fiber and acid detergent fiber was lower in young leaves than in mature.

Finally, the leaves of *Q. rugosa* can be included in diets for maintenance of sheep up to 30%, without negative effects on palatability and consumption. Moreover, the contents of organic matter, neutral detergent fiber and acid detergent fiber decreased as molasses levels were increased in mixtures (Quercus + molasses + urea). However, the CP, non-fiber carbohydrates, ash, metabolizable energy, short chain fatty acids and net energy for lactancy had an opposite behavior. The ether extract and hemicellulose were reduced with increasing the urea inclusion in diets. In general, the *in vitro* digestibility of organic matter increased as molasses in mixtures augmented. In all mixtures of molasses, fractions **b** and **a+b** were higher with 4 or 6% of urea. However, the constant rate **c** of degradation varied between treatments.

In Tables 40.3, 40.4 and 40.5 are listed the nutritional characteristics of leaves of *Quercus eduardii*, consumed by range ruminates and collected in summer at the north region of Durango State, Mexico. It has low crude protein medium lignin content, higher hemicellulose than cellulose, very low values of condensed tannins and ether extract, the acetic acid is much higher than propionic or butyric, isobutyric, isovaleric valeric acids. Except of Ca, Fe and Mn, all minerals were lower to fulfill the adult small ruminant metabolic requirements. The *in vitro* gas production was low along with the bypass protein, microbial protein and metabolizable protein. The potential dry matter digestibility was also low; in addition, it has low the potential digestible and metabolizable energy and medium values of dry matter intake of range ruminant consuming leaves of *Quercus eduardii*.

In Tables 40.6, 40.7 and 40.8 are listed the nutritional characteristics of leaves of *Quercus grisea*, consumed by range ruminates and collected in summer at the north region of Durango State, Mexico. It has low crude protein medium lignin content, higher hemicellulose than cellulose, very low values of condensed tannins and ether extract, the acetic acid is much higher than propionic or butyric, isobutyric, isovaleric valeric acids.

Except of Na, Cu and Zn, all minerals were sufficient to fulfill the adult small ruminant metabolic requirements. The *in vitro* gas production was low along with the bypass protein, microbial protein and metabolizable protein. The potential dry matter digestibility was also low; Moreover, there were low the potential digestible and metabolizable energy and medium values of dry matter intake of range ruminant consuming leaves of *Quercus grisea*.

Table 40.3. Chemical composition (%) and *in vitro* volatile fatty acids (mM) of leaves from *Quercus eduardii* collected in the state of Durango México

Concept	%, dry mater
Ash	7
Crude protein	7
Neutral detergent fiber	64
Acid detergent fiber	34
Lignin	10
Hemicellulose	31
Cellulose	23
Condensed tannins	1
Ether extract	1
Acetic mM	14
Propionic	2
Butyric	2
Isobutyric	0.1
Isovaleric	0.2
Valeric	0.4
Total	19
Acetic:Propinic	7

Table 40.4. Mineral composition of macro (g/kg, dry matter) and microminerals (mg/kg, dry matter)of leaves from *Quercus eduardii* collected in the state of Durango México

Concept	Minerals
Macrominerals	
Ca	8
K	1
Mg	1
Na	1
P	1
Microminerals	
Cu	2
Fe	208
Mn	48
Zn	16

Table 40.5. *In vitro* gas production in 24 hours (GasP$_{24h}$/ mL 200 mg, dry matter), bypass protein (%, dry matter), microbial protein (%, dry matter) and metabolizable protein (%, dry matter) in leaves from *Quercus eduardii* collected in north of the state of Durango México

Concept	Dry matter
In vitro gas production	13
Bypass protein	3
Microbial protein	1
Metabolizable protein	5
Potential dry matter digestibility	57
Potential digestible energy, Mcal/kg	2.7
Potential metabolizable energy, Mcal/kg	2.2
Potential dry matter intake, g/kg/day	81

Table 40.6. Chemical composition (%) and *in vitro* volatile fatty acids (mM) of leaves from *Quercus grisea* collected in the state of Durango México

Concept	Dry mater basis
Ash	5
Crude protein	9
Neutral detergent fiber	65
Acid detergent fiber	32
Lignin	10
Hemicellulose	33
Cellulose	21
Condensed tannins	0
Ether extract	1
Acetic	9
Propionic	1
Butyric	1
Isobutyric	0.3
Isovaleric	0.3
Valeric	0.5
Total	13
Acetic:Propinic	7

Table 40.7. Mineral composition of macro (g/kg, dry matter) and microminerals (mg/kg, dry matter)of leaves from *Quercus grisea* collected in the state of Durango México

Concept	Minerals
Macrominerals	
Ca	11
K	4
Mg	2
Na	1
P	2

Microminerals	
Cu	4
Fe	63
Mn	54
Zn	16

Table 40.8. *In vitro* gas production in 24 hours (GasP$_{24h}$/ mL 200 mg, dry matter), bypass protein (5, dry matter), microbial protein (%, dry matter) and metabolizable protein (%, dry matter) in leaves from *Quercus grisea* collected in north of the state of Durango México

Concept	Dry matter
In vitro gas production	15
Bypass protein	3.2
Microbial protein	2.1
Metabolizable protein	5.4
Potential dry matter digestibility	59
Potential digestible energy, Mcal/kg	2.8
Potential metabolizable energy, Mcal/kg	2.3
Potential dry matter intake, g/kg/day	81

CHAPTER 41

Nutrition of *Schaefferia cuneifolia* A. Gray.

Introduction

Schaefferia cuneifolia belongs to the family Celastraceae is commonly known in Mexico as capull. In the US as desert yaupon. Its fruits are eaten by birds, is browsed by sheep, goats and cattle, and occasionally is browsed by white-tailed deer. Evergreen shrub about 3 to 5 m; leaves small and green, usually obovate; flowers light green to white. The berries turn red when developed, eaten by birds and small wild mammals. The taxonomic characteristics of *Schaefferia cuneifolia* are listed in Table 41.1.

Schaefferia cuneifolia

Table 41.1. Taxonomic characteristics of *Schaefferia cuneifolia*

Rank	Scientific Name and Common Name
Kingdom	Plantae – Plants
Subkingdom	Tracheobionta – Vascular plants
Superdivision	Spermatophyta – Seed plants
Division	Magnoliophyta – Flowering plants
Class	Magnoliopsida – Dicotyledons
Subclass	Rosidae
Order	Celastrales
Family	Celastraceae – Bittersweet family
Genus	*Schaefferia* Jacq. – Schaefferia
Species	*Schaefferia cuneifolia* A. Gray – desert yaupon

Nutritional value

In Tables 41.2, 41.3 and 41.4 are listed the nutritional characteristics of Schaefferia cuneifolia collected in northeastern Mexico. It has high organic matter and crude protein especially in summer and fall. Low neutral detergent fiber content. The acid detergent fiber is mostly composed by cellulase and lignin. The acid detergent fiber (ADF) is the indigestible fraction of plant material in forage, usually cellulose fiber coated with lignin. It is generally assumed that the digestion of a food is inversely proportional to its ADF content; thus, ADF is utilized to compute energy values of feedstuffs. Nevertheless, the relationship between ADF and digestibility is sometimes poor, ADF frequently accounting for less than 65% of the variance in digestibility ($r^2 < 0.65$). In addition, in some cases, particularly for second cut forages, the relationship can be positive.

The predicted values of dry matter digestibility, digestible energy, metabolizable energy and dry matter intake were sufficient for the maintenance needs of adult small ruminants. The concentrations of Ca, K, Mg, Cu, Fe, Mn and Zn were sufficient to satisfy the maintenance requirements of small ruminants. However, Na and P were insufficient.

Table 41.2. Seasonal variations of the chemical composition (%, dry matter) of leaves from *Schaefferia cuneifolia* collected in northeastern Mexico

Concept	Seasons				Annual
	Winter	Spring	Summer	Fall	mean
Organic matter	79	83	86	81	82
Ash	21	17	14	19	18
Crude protein	12	15	18	14	15
Degradable protein	10	12	15	11	12
Undegraded protein	2	3	3	3	3
Neutral detergent fiber	51	48	38	41	44
Insoluble neutral detergent fiber	12	11	9	9	10
Cellular content	49	52	62	49	56
Acid detergent fiber	32	29	25	24	27
Cellulose	23	9	12	17	15
Hemicellulose	19	19	12	17	17
Lignin	13	11	9	12	12
Condensed tannins	0	0	0	0	0
Gross energy, Mcal/kg	4	4	4	4	4

Table 41.3. Seasonal variation of predicted dry mater digestibility (%), digestible energy (Mcal/kg), metabolizable energy (Mcal/kg) and dry matter intake (g/kg BW/day) of the leaves of *Schaefferia cuneifolia* collected in northeastern Mexico

Concept	Seasons				Annual
	Winter	Spring	Summer	Fall	mean
Dry matter digestibility	65	62	60	66	64
Digestible energy	3.0	2.9	2.8	3.1	3.0
Metabolizable energy	2.5	2.4	2.3	2.5	2.5
Dry matter intake	83	82	82	83	83

Table 41.4. Seasonal content of macro (g/kg, dry matter) and microminerals (mg/kg, dry matter) in leaves of *Schaefferia cuneifolia* collected in northeastern Mexico

Concept	Seasons				Annual mean
	Winter	Spring	Summer	Fall	
Macrominerals					
Ca	60	56	66	59	64
K	15	10	15	16	14
Mg	7	3	5	5	5
Na	<1	<1	<1	<1	<1
P	1	1	1	1	1
Microminerals					
Cu	10	11	13	10	11
Fe	125	120	139	130	127
Mn	30	35	36	37	35
Zn	39	38	51	31	40

CHAPTER 42

Nutrition of *Zanthoxylum fagara* (L.) Sarg.

Introduction

Zanthoxylum fagara is a species of the citrus family, Rutaceae. It offers important feed and moderate quantities of cover for wildlife. It has a high dry tolerance and grows greatest in full sun, but it may also subsist as an understory browse plant. It offers important feed and cover for white-tailed deer. It is a shrub or small tree reaching a height of 2-12 m high. Most species are trees or shrubs, a few are herbs, commonly perfumed with leaf glands, sometimes with spines. The leaves are frequently opposite and compound, and lacking stipules. The taxonomic characteristics of *Zanthoxylum fagara* are shown in Table 42.1.

Zanthoxylum fagara

Table 42.1. Taxonomic characteristics of *Zanthoxylum fagara*

Rank	Scientific Name and Common Name
Kingdom	Plantae – Plants
Subkingdom	Tracheobionta – Vascular plants
Superdivision	Spermatophyta – Seed plants
Division	Magnoliophyta – Flowering plants
Class	Magnoliopsida – Dicotyledons
Subclass	Rosidae
Order	Sapindales
Family	Rutaceae - Rue family
Genus	*Zanthoxylum* L. – pricklyash
Species	*Zanthoxylum fagara* (L.) Sarg. – lime pricklyash

Nutritional value

In Tables 42.2, 42.3 and 42.4 are listed the nutritional characteristics of Zanthoxylum fagara collected in northeastern Mexico. In general, it has high organic matter, crude protein and degradable protein content. It has low neutral detergent fiber and hemicellulose content is higher than cellulose or lignin content. It does not has condensed tannins and the gross energy is similar than other evaluated shrubs. The predicted values of dry matter digestibility, digestible energy, metabolizable energy and dry matter intake appeared to be good enough to satisfy the metabolic needs of adult range small ruminants.

The concentration of Na, P and Cu were marginal lower for the metabolic requirements of adult range small ruminants. However, the content of Ca, K, Mg, Fe, Mn and Zn were sufficient. Fodders are good sources of minerals for range ruminants. In some situations, they may provision adequate quantities of all essential minerals required for ruminants. Nonetheless, in distinct situations fodders are deficient in one or more minerals and are required to be complemented for optimal animal performance or health. The unbalances of minerals (excess or deficiencies) in soils and forages has been considered responsible for low ruminant efficiency in browsing ruminants.

The capability of fodders to give the ruminant with satisfactory source of minerals is determined by the bioavailability and mineral content. The bioavailability is well defined as the quantity of the mineral that is consumed, conveyed to the action place and transformed in a physiological form. An important characteristic for mineral absorption is that the mineral has to be soluble. Minerals may be transported in a ionic form, as soluble complex or chelated, depending of a specific element; however, minerals cannot be absorbed when are in the insoluble form.

Table 42.2. Seasonal variations of chemical composition (%, dry matter) and gross energy in leaves of *Zanthoxylum fagara* collected in northeastern Mexico

Concept	Seasons				Annual
	Winter	Spring	Summer	Fall	mean
Organic matter	86	89	88	87	88
Ash	14	11	12	13	12
Crude protein	20	26	21	19	21
Degradable protein	17	17	15	13	15
Undegraded protein	3	9	6	6	6
Neutral detergent fiber	26	40	33	26	32
Insoluble neutral detergent fiber	7	8	6	6	7
Cellular content	74	60	67	74	68
Acid detergent fiber	18	22	16	16	18
Cellulose	9	11	7	7	9
Hemicellulose	8	18	17	11	14
Lignin	8	10	8	8	8
Condensed tannins	0	0	0	0	0
Gross energy, Mcal/kg	5	5	5	5	5

Table 42.3. Seasonal variation of predicted dry mater digestibility (%), digestible energy (Mcal/kg), metabolizable energy (Mcal/kg) and dry matter intake (g/kg BW/day) of the leaves of *Schaefferia cuneifolia* collected in northeastern Mexico

Concept	Seasons				Annual mean
	Winter	Spring	Summer	Fall	
Dry matter digestibility	72	70	74	73	72
Digestible energy	3.3	3.2	3.4	3.4	3.4
Metabolizable energy	2.7	2.7	2.8	2.8	2.8
Dry matter intake	84	83	84	84	84

Table 42.4. Seasonal content of macro (g/kg, dry matter) and microminerals (mg/kg, dry matter) in leaves of *Zanthoxylum fagara* collected in northeastern Mexico

Concept	Seasons				Annual mean
	Winter	Spring	Summer	Fall	
Macrominerals					
Ca	48	38	48	47	45
K	13	14	18	12	14
Mg	6	4	5	5	5
Na	<1	<1	<1	<1	<1
P	1	2	2	2	2
Microminerals					
Cu	7	5	6	5	6
Fe	156	164	254	218	198
Mn	103	81	132	170	122
Zn	32	57	39	39	42

CHAPTER 43

Nutrition of *Ziziphus Obtusifolia* (T & G)

Introduction

Ziziphus obtusifolia belongs to the family Rhamnaceae. It is known as lotebush. The foliage is browsed by goats, cattle, sheep and white-tailed deer, but with low preference. It is considered as of medium food value for white-tailed deer. *In vitro* dry matter digestion is about 45.4%. Leaf crude protein ranged from 19% in May to 40% in March. Nutritional value (as percentage of dry weight) for lotebush fruits is as follows: crude protein 11, P 0.19, Ca 0.35, Mg 0.12, K 1.23 and Na 0.04. The taxonomic characteristics of *Ziziphus Obtusifolia* collected in northeastern Mexico are listed in Table 43.1.

Ziziphus Obtusifolia

Table 43.1. Taxonomic characteristics of *Ziziphus Obtusifolia*

Rank	Scientific Name and Common Name
Kingdom	Plantae – Plants
Subkingdom	Tracheobionta – Vascular plants
Superdivision	Spermatophyta – Seed plants
Division	Magnoliophyta – Flowering plants
Class	Magnoliopsida – Dicotyledons
Subclass	Rosidae
Order	Rhamnales
Family	Rhamnaceae – Buckthorn family
Genus	*Ziziphus* Mill. – jujube
Species	*Ziziphus obtusifolia* (Hook. ex Torr. & A. Gray) A. Gray – lotebush

Nutritional value

The nutritional profile of leaf *Ziziphus obtusifolia*, collected in northeastern Mexico is shown in Table 43.2, 43.3 and 43.4. In general, it has high organic matter, crude protein and degradable protein content. It has low neutral detergent fiber. The hemicellulose and lignin are very similar, but both are higher than cellulose. The leaves have high-condensed tannins content. The acetic acid is more than five times greater than propionic or butyric acids. The predicted values of dry matter digestibility, digestible energy, metabolizable energy and dry matter intake were sufficient for the metabolic needs of adult range small ruminants.

The concentrations of Na, P and Zn were insufficient for the metabolic requirements of adult range small ruminants. However, Ca, K, Mg, Cu, Fe and Mn contents were sufficient. One of the central problems that reduce the animal performance in many regions of the world is the nutritional status of ruminants. In agreement with some scientific reports, the main characteristics that diminish productivity of range animals are: 1) low protein content in native fodders, 2) low energy consumption because of high cell wall content in range fodders and 3) lack of minerals and/or vitamins. In addition, it has to take account that the problems related to mineral nutrition not only are because of deficiencies, but also with toxic minerals such as Hg, Al, Cd, Pb, even with essential minerals such as Cu, Fe, M), and Se. Moreover, Ca, Cu or Se in excess may cause negative effects on ruminant performance.

Table 43.2. Seasonal variations of the chemical composition (%, dry matter) and molar proportions (Mol/100 Mol, dry matter) in leaves of *Ziziphus obtusifolia* collected in northeastern Mexico

Concept	Seasons				Annual
	Winter	Spring	Summer	Fall	mean
Organic matter	94	91	90	91	90
Ash	6	9	10	9	9
Crude protein	19	16	14	16	16
Degradable protein	12	10	9	10	10
Undegraded protein	5	6	5	6	6
Neutral detergent fiber	28	26	30	28	28
Insoluble neutral detergent fiber	7	6	6	7	7
Cellular content	72	74	70	72	72
Acid detergent fiber	17	15	14	17	17
Cellulose	5	6	5	5	5
Hemicellulose	11	9	16	11	11
Lignin	12	11	12	11	12
Condensed tannins	13	14	11	14	14
Acetic acid	81	78	83	81	80
Propionic acid	11	16	7	12	12
Butyric acid	8	6	10	7	8

Table 43.3. Seasonal variation of predicted dry mater digestibility (%), digestible energy (Mcal/kg), metabolizable energy (Mcal/kg) and dry matter intake (g/kg BW/day) of the leaves of *Ziziphus Obtusifolia* collected in northeastern Mexico

Concept	Seasons				Annual
	Winter	Spring	Summer	Fall	mean
Dry matter digestibility	73	74	74	72	72
Digestible energy	3.4	3.4	3.4	3.4	3.4
Metabolizable energy	2.8	2.8	2.8	2.8	2.8
Dry matter intake	84	84	84	84	84

Table 43.4. Seasonal content of macro (g/kg, dry matter) and microminerals (mg/kg, dry matter) in leaves of *Ziziphus obtusifolia* collected in northeastern Mexico

Concept	Seasons				Annual mean
	Winter	Spring	Summer	Fall	
Macrominerals					
Ca	8	9	8	7	9
K	22	21	25	23	23
Mg	9	7	7	6	7
Na	<1	<1	<1	<1	<1
P	1	1	1	1	1
Microminerals					
Cu	19	17	18	13	17
Fe	93	67	112	84	89
Mn	60	65	59	64	62
Zn	20	17	18	14	18

CHAPTER 44

Diet selection by ruminants

Introduction

Each species has very different dietary habits and preferences that are molded from their environment, their genetics, and their social interactions. In general, animals consume foods that they are physiologically adapted to digest and that meet their nutritional requirements. These inherent dietary differences result in herbivores being classified into three major groups: grazers, browsers, and intermediate feeders. In addition, physiology alone does not dictate diet selection in animals. The diets of animals are strongly influenced by 1) social interactions with mother, peers, and people; 2) feedback from nutrients and toxins in plants; and 3) interactions with their physical environment including location of water and predators. Animals grazing mixed-species rangelands face numerous selections, as well as where and when to ingest and how abundant foliage to ingest. These selections disturb not only the nutritive standing of the ruminant, but also the grassland structure and nutritious value through discerning selection of herbage.

Digestive physiology of browsers

The gastro intestinal track of many range ruminants permits them to obtain nutrients from an extensive diversity of plants. Nevertheless, ruminants that are most effective when finding necessary nutrients are those that can be more probable to live, replicate, and yield products like milk, skin, and meat. When calculating the selected diets by ruminants leads to improved range management and habitat, and permits good empathetic of connections between domestic and wild ruminants. Thus, the association among morphology and nutrient extraction in ruminants is an investigation of importance.

Ruminants have a digestive system intended to ferment foods and offer predecessors for energetic compounds for the body tissues. They differ from nonruminants due to they have four sections in the front section of their digestion tract and due to ruminants masticate their bolus. The bolus is in due course regurgitated and chewed to entirely mix it with saliva and to break the particles into small size. The abomasum is similar as the stomach in the nonruminant. The rumen generates a physical constraint to the passage of feed through the gastro intestinal digestive tract. For ingesta to leave the rumen, the feed particles have to be minor and substantial that needs masticate again and time to ferment in the rumen. About 155 ruminant species are all over the world. Most large herbivores on semiarid rangelands of northeastern Mexico are ruminants such as cattle, sheep, goats, and deer.

Ruminants have stomach with four sections. The compartments are rumen, reticulum, omasum and abomasum. In the rumen and the reticulum, the feedstuffs are mixed with saliva and splits into strata of liquid and solid resources. The solids bunch together to form the bolus. The dietary carbohydrates hemicellulose and cellulose, are mainly fragmented in rumen by microorganisms (bacteria, protozoa, fungi and yeast) into the volatile fatty acids (VFA): acetic acid, propionic acid, and butyric acid. In addition, N compounds and soluble carbohydrates such as pectin and sugars are fermented.

Even though the reticulum and rumen have different names, they denote similar useful space as ingesta may move back and forth among them. These two sections are named the reticulorumen. The break down ingesta that is now in the bottom liquid section of the reticulorumen eventually passes to the following section, the omasum, where many of the mineral elements and water are immersed into the blood torrent. After that, the ingesta is progressed to the abomasum. The ingesta is finally passed to the small intestine, where the nutrients are digested and absorbed. Microorganisms formed in the reticulorumen are also digested in the lower track. The fermentation remains in the lower track in the similar manner like in the reticulorumen.

Many dietary carbohydrates are converted into VFA in the rumen. Only small amounts of dietary glucose are absorbed. The glucose that is requited for energy for the brain and for the disaccharide <u>lactose</u> and for milk yield, originates from the nonsugar sources, such as the VFA propionic acid, proteins, lactic acid and glycerol. The propionic acid is used in about 70% of the glucose and <u>glycogen</u> produced and 20% for protein.

Browser ruminants

Browser ruminants consume mostly browse and non-browse dicotyledonous foliage, for example trees, shrubs, herbs, or fruits. Most browses have low plant cell wall and fibers in their cell wall are additional indigestible and lignified, thus the smaller rumen of browsing animals may permit indigestible feed elements to movement more fast through the digestive tract. This quick movement may endorse a higher feed consumption. Browsers incline to have widespread condensed papillae in all sections of the rumen, expanding the surface part by 22 times that van permit effective absorption of VFA from the quickly digested cell contents of browse materials. Browsers have a respectively big abomasum, a larger cecum and colon, and the ventricular groove in the rumen/reticulum can permit some cell contents to avoid incompetent rumen digestion in favor of direct fermentation in the abomasum and lower intestinal digestive system.

The parotid salivary glands of browsers yield salivary proteins that bind tannin, which can avoid tannins in browses from significantly diminish protein digestion. It also has been noticed that browsers have much more liver tissue for their body extent than grazers do.
Browsers incline to have a narrower muzzle and a comparatively larger mouth opening that allows indirect stripping of leaves.

Classification of ruminants in relation to their dietary habits

The most broadly recognized practical organization of ruminants related to their nutritional habits, is that classifies them into browsers, intermediate feeders and grassers. This arrangement is founded on the quantity of diverse fodders (browse or grasses) that the ruminants comprise in their diets. Animals that consume mainly on parts of browse plants are recorded as browsers. However, grazing animals are measured to be those that eat primarily grasses. Intermediate feeders are a mixed group of species with diets that contains a combination of both grasses and browse.

Grazers

Grazers, including horses and cattle, mainly eat grasses and have the digestive abilities of handling large amounts of roughages. The sheer size of their mouth, and their overall size, cattle are better improved to select grass than browse plants. Cattle have a great muzzle and lips and a tongue that is utilized like prehensile searching device. The great muzzle restrictions their capability to choice among plants and the different parts of a plant. They select using their tongue to trap plants into their mouth cavity where the ingesta is strained between the upper dental pad and lower incisors and then torn off. Because the rumen of cattle is large, it gives the ability to digest low quality roughage, which makes them superior to sheep or goats when consuming fibrous and plentiful plant species. Examples of grassers are: cattle, musk oxen, bison (usually large ungulates).

Browsers

Browsers like deer emphasis their diet choice, first on the leaves, flowers, and twigs of woody species. They naturally have a reduced, sharper mouth than grazers. The dental adaptations of browsers, narrowed muzzle benefit animals to select separate parts of a plant of higher nutritional profile. Generally, browsing animals select diets with higher crude protein and digestibility than those diets of grazing animals. However, many woody plants have toxic secondary compounds that diminish their consumption by browsers. Browsers have established some physiological features that assistance them either digest or evade contact to these complexes. For instance, the liver of several browsers have is larger relative to their body dimensions that helps in the metabolism of plant poisons. Some browsing herbivores are armed with salivary glands that bind tannins, an antinutritional compound found in some woody species. Examples of browsers are: deer, giraffe, rabbit.

Intermediate Feeders

Intermediate feeders have adaptations to be browsers and grazers. They have a comparatively small mouth permitting them to graze quite close to the ground and to shred flowers or leaves from stems. Their diet selection usually is conquered by forbs, though they might have voluntarily intake grasses when grass plants are moist or when other fodder has inadequate accessibility. Examples of intermediate feeders are: sheep, goats, elk (usually medium sized ungulates)

Competition between ruminant types

Range ruminants vary extensively in the classes of fodders that they are adapted to utilize. These modifications are manly founded on the body anatomy of the ruminants. In ruminants, the grade to which a ruminant can acclimatize to diverse diets is associated to its feeding habit than is resulted by its gastrointestinal structure. The least-adaptable animals are the grazers and browsers. Between these two sets are the intermediate feeders that are tremendously plastic in their diets and, consequently, the habitats they may use. Even though grazers will consume browse and browsers consume grasses, they will not achieve well when are enforced to change their diets to that excesses. Landowner's can improve the ability to successfully manage different range herbivores, if they understand the variations in feeding categories and what feed sources are appropriate for the ruminants.

Characteristics that affect diet selection

Animals eat feeds that they are physiologically adjusted to metabolize and that fulfill their nutritional needs. Nonetheless, the capability to digest feedstuffs and the nutritional needs of each animal can differ significantly depending on their reproductive condition, body status, age and gender. Goats are bodily alert animals that may position on their hind legs to extent tall growing browse foliage or use their forefeet to pull down the branches to bit leaf materials. Shorter goats may even climb trees to gain entree to upper foliage. Their physical nature permits goats to holder coarser and sharper territory than sheep or cattle.

Selecting an animal type.- Breeds of livestock fluctuate in body size and performance features that command their nutrient needs, dry matter consumption, and digestion capability. These features affect which plants, and in what quantity, of animals select to comprise in their diets.

Multi species browsing.- Invasive plant species can be eliminated selecting the livestock species, which more rapidly eats the plants designed for eradicate. When using more than one animal species can improve the profits. That grazing utilizes two or more species to graze the same portion of grassland, but not necessarily at the similar period. It is possible to reestablish equilibrium to systems by encouraging extra application of all fodder species, avoiding an ecological benefit for each species of plants or plant classes.

Age of the animal.- The age deeply upset diet selection and acceptance to plant toxic compounds. In addition, metabolic needs vary with age thus; older animals require fewer foods and employ less time browsing. Conversely, adults, young, growing animals require rations higher in energy and protein and with less fiber. Their pursuit for an extra nutritious ration requires more energy. This, together with incomplete information, can lead younger animals to attempt original feeds and repeat foods that once made them sick. For instance, younger animals seem more enthusiastic than older animals to intake less needed fodders. Those just weaned are increasing their diet selections, so they are also more enthusiastic to attempt new feeds.

Animal sex.- Females and males choice diverse diets, This is due to differences in frame and complete nutrient needs during the reproductive period. Morphological and physiological characteristics, like development rate and food adaptation efficacy, also affect to modifications in diets. Males usually have superior height and nose size than females and can have superior energy requirements. Changes in browsing performance among females and males are extensively familiar but not well understood.

Production status.- Cycle animals select their diets founded on nutritional requirements that modify intensely in life periods. This information may aid with prescribed browsing. For instance, some aggressive plants with elevated nutrient profile can fulfill the needs of females on lactation and growing newborns. Revisions specify that sheep grazing green spurge wean substantial lambs than their complements grazing spurge on free range. Nevertheless, not all aggressive plants are very nutritious, and animals have to have sufficient substitute fodder to keep productivity before reproduction to come across the nutrition requirements during lactation and gestation.

Animal corporal condition.- Depending how thin or fat an animal is effects its browsing conduct. Those in low frame condition or on in a ration insufficient to meet their maintenance needs can have less tolerance for plant poisons. This is due to there is a nutritional price to metabolize a poisons or aversive plant complex. In the liver most often occurs the detoxification, thus; an animal that intakes toxic plants requires a big, vigorous liver. Continued nutritional stress may diminish liver mass. Mineral and protein complements may improve the function of microbes in the rumen, hepatic enzymes, and complexes for conjugating poisons, all of them improve the abilities of an animal for detoxification.

Factors inherited that affect diet selection in arid conditions

Browsing is a practice by which range ruminants discovery and ingest the food for living. So far it has only be debated how ruminants make selections after they discovery food. Nonetheless, browsing also needs a moment of rambling about the environmental area to discovery these supplies. The rambling aptitudes of ruminants are at least hereditary. The aptitude of the ruminant to handle sharp topography, browse or grass in regions without shadow, or transport excessive spaces from water has been demonstrated to disturb diet selectivity in domestic ruminants. Under desert range conditions, alterations in diets selectivity by range cattle were accredited in part, to how distant ruminants moved from water. This idea applies similarly well to wildlife ruminants.

Selective grazing/browsing

This reciprocal process regulates the nutritious wellbeing of the ruminant, and instead, modifies the changing aspects of the vegetation community. Thus, it is vital to comprehend how ruminants select while grazing/browsing. Current diet selectivity concepts recommend that feed favorites and dislikes be founded on knowledges within the life span of the ruminant. The dietary preferences and aversions of grazing/browsing ruminants are definitely learned actions, but hereditary morphological, biological, and neural features may modify the nature and scale of digestive response. Thus, diet predilections could be hereditarily passed from parentages to young animals. Accepting the inheritance selectivity of diets might help leaders progress the biological sustainability of grazing/browsing livestock. The selectivity and breeding of ruminants with precise dietary features might also be utilized to create flocks and herds of livestock to regulate wild plants or succeed wildlife habitat with treatment grazing/browsing systems. Thus, selective breeding of dietary features might establish a dominant new range administration instrument.

Optimum habitat condition

It is a concept characteristically utilized for wildlife conditions instead for livestock. The description for optimal livestock habitat would vary with organization purposes. Abiotic aspects, like landscape, water accessibility, and thermal shelter, disturb animal routine and regularity of grazing/browsing. Livestock frequently desire moderate slopes and evade roaming extended horizontal and vertical travels to water. Shadow and close water are utilized for thermoregulation when temperatures are elevated and topographic break and browse vegetation may be utilized for warm cover in chiller temperatures. Biotic aspects, such as foliage quantity and quality, effect spatial grazing/browsing predilections and disturb animal productivity. Livestock desire regions with higher foliage quantity and quality. Homogeneity of grazing/browsing may be better in similar vegetation, but animal productivity may be better in mixed vegetation, particularly at lower stocking rates. Livestock grazing/browsing arrangements have been expected using multiple regression analyses and other predictive models; however, their accomplishment has characteristically been partial to a precise site. Administrators may expand livestock habitat situations by varying abiotic qualities of the meadows, such as emerging water, construction arrangements for thermal cover, and varying biotic qualities of the grassland by burning, manuring, changing stocking rates, and handling grazing/browsing systems. Administrators may in addition, select animals that are more adjusted to precise rangeland aspects. Performs such as planned complementation and herding can modify livestock behavior outlines to use more of the accessible environment. The spatio-temporal changeability of rangeland needs numerous management applications to improve the procedure of animal habitat.

CHAPTER 45

References

Ammar H., López S., Bochi O., Garcia R., Ranilla M.J., 1999. Composition and in vitro digestibility of leaves and stems of grasses and legumes harvested from permanent mountain meadows at different maturity stages. J. Anim. Feed Sci. 8: 599-610.

Ammar H., López S., González J.S., Ranilla M.J., 2004188. Comparison between analytical methods and biological assays for the assessment of tannin-related antinutritive effects in some Spanish browse species. J. Sci. Food Agr. 84: 1349-1356.

Beverly, C.A., Goff C.M., T. Forbes, D.A. 1998. Toxic amines and alkaloids from Acacia rigidula, Phytochemistry Vol. 49: 1377-1380.

Juarez, R.A.S., C.G. Nevarez, H.C.A, Meza, and S.M.A. Cerrillo. (2004). Diet composition, intake, plasma metabolites, reproductive, and metabolic hormones during pregnancy in goats under semiarid grazing conditions. J. Agric. Sci. 142(6):697-704.

Abdulrazak, S.A., Fujihara, T., Ondiek, J.K., Ørskov, E.R. 2000. Nutritive evaluation of some acacia tree leaves from Kenya. Animal Feed Science and Technology, 85(1-2): 89-98.

Abdul-Razzaq, H. A., R. Bickerstaffe and G. P. Savage. 1988. The influence of rumen volatile fatty acid on blood metabolites and body composition of growing lambs. Aust. J. Agric. Res. 39:505-515.

Acero, A., J.P. Muir, and R.M. Wolfe (2010). Nutritional composition and condensed tannin concentration changes as browse leaves become litter. J. Science Food Agr. 90:2582-2595.

Ackerman, B. A. 1983. Las Gramíneas de México. Tomo I. SARH-COTECOCA. México, D. F. pp. 259-268.

Adesogan, A.T. (2005). Effect of bag type on the apparent digestibility of feeds in ANKOM DaisyII incubators. Anim. Feed Sci. Technol. 119:333-344.

Aganga, A.A., Tshwenyane, S.O. 2003. Feeding Values and Anti - Nutritive Factors of Forage Tree Legumes. Pakistan Journal of Nutrition 2: 170-177.

Aguilera, S. J. I. 1998. Perfil de la concentración de ácidos grasos volátiles de diez arbustivas nativas del noreste de México. Tesis de Licenciatura. Facultad de Medicina Veterinaria y Zootecnia, UAZ. El Cordovel, Zacatecas.

Agundez, J. E., M. H. Fraga, H. A. Escobar y T. A. Gillen. 1993. Especies forrajeras silvestres en la región de los cabos, BCS. Revista de Investigación Científica, UABCS, México. 3:1

Alanís F., G. J. & S. Favela L. 1997. Plantas nativas usadas en el árido paisaje (jardines xéricos). Memorias del Primer Congreso Nacional para el aprovechamiento integral de recursos de zonas áridas. Unidad Regional Universitaria de Zonas Áridas, Universidad Autónoma de Chapingo, Bermjillo, Durango.

Alanis-Flores, G.J., G. Cano y Cano y M. Rovalo Merino. 1996. Vegetación y Flora de Nuevo León, una guía Botánico-Ecológica. Patronato Monterrey 400, Consejo Consultivo para la Preservación y Fomento de la Flora y Fauna Silvestre de Nuevo León, CEMEX. Monterrey, Nuevo León, México.

Al Jassim, R.A.M., K.I. Ereifej, R.A. Shibli y A. Abudabos. 1998. Utilization of concentrate diets containing acorns (*Quercus aegilops* and *Quercus coccifera*) and urea by growing Awassi lambs. Small Rum. Res. 29 (3), pg. 289-293.

Al-Soqeer A.A., 2008. Nutritive value assessment of Acacia species using their chemical analyses and in vitro gas production technique. Res. J. Agr. Biol. Sci. 4: 688-694.

Amo Silvia del R. 1979. Plantas medicinales del estado de Veracruz. Instituto Nacional de Investigaciones sobre Recursos Bióticos. Xalapa, Veracruz. P. 119.

Ansley, R. J.; Jacoby, P. W.; Lawrence, B. K. 1989. Influence of stress history on water use patterns of honey mesquite. In: Wallace, Arthur; McArthur, E. Durant; Haferkamp, Marshall R., compilers. Proceedings--symposium on shrub ecophysiology and biotechnology; 1987. June 30 - July 2; Logan, UT. Gen. Tech. Rep. INT-256. Ogden, UT: U.S. Department of Agriculture, Forest Service, Intermountain Research Station: 75-82.

Ansley, R. J.; Jacoby, P. W.; Cuomo, G. J. 1990. Water relations of honey mesquite following severing of lateral roots: influence of location and amount of subsurface water. Journal of Range Management. 43(5): 436-442.

AOAC. (1997). Official Methods of Analysis. Association of Official Analytical Chemists, (Gaithersburg, Maryland, USA).

Archer, Steve. 1990. Development and stability of grass/woody mosaics in a subtropical savanna parkland, Texas, U.S.A. Journal of Biogeography. 17: 453-462.

Aregawi, T., Melaku, S., Nigatu, L. 2008. Management and utilization of browse species as livestock feed in semi-ariddistrict of North Ethiopia, vol 20, Article #86. Livestock Researchfor Rural Development,

Armenta-Quintana, J.A., Ramírez, R.G., Ramírez-Orduña, R. 2009. Organic matter degradability of diets by range goats. Journal of Animal and Veterinary Advances, 8(5): 825-828.

Arteaga, S.; Andrade-Cetto, A.; Cardenas, R. (2005). "Larrea tridentata (Creosote Bush), an abundant plant of Mexican and US-American deserts and its metabolite nordihydroguaiaretic acid". Journal of Ethnopharmacology, 98(3): 231–239.

Ash, A.J. 1990. The effect of suplementation with leaves from the leguminous trees Sesbania grandiflora, Albizia chinensis and Gliricia seprum on the intake and digestibility of Guinea Grass Hay by Goats. Anim. Feed. Sci. and tech. 28:225-232.

Asleson, M.A., E.C. Hellgren, and L. W. Varner. 1996. Nitrogen requirements for antler growth and maintenance in white-tailed deer. Journal of Wildlife Management 60: 744-752.

Attaie-R; Ritcher-RL; Reine-A. H. 1992. Volatile branched chain and n-chain fatty acid in caprine and bovine colostrum. Journal-of-Dairy-Science. 1992, 75:Suplement 1, 119.

Avila C., J.M., J.A. Ortega S. y A. Flores M. 1992. Efecto de la escarificación con agua en la germinación de semillas de Desmanthus virgatus. Reunión Nac. De Invest. Pec. En México. INIFAP-UNAM-UACH. Chih., México, p. 26.

Barnes, T. G., L. W. Varner, L. H. Blankenship, J. F. Gallagher. 1991b. Indigestible particulate passage in white-tailed deer. In: R. D. Brown [ed.]. The biology of deer. New York, NY: Academic Press, p 436-442.

Batos, B., Z. Miletic, S. Orlovic y D. Miljkovic. 2010. Variability of Nutritive Macroelements in Pedunculate Oak (Quercus robur L.) Leaves in Serbia. Genetika-Belgrade. 42, 435-453.

Beauchemin, K.A., and W.Z. Yang. 2005. Effects of physical effective fiber intake, chewing activity, and ruminal acidosis for diary cows fed diets based on corn silage. J. Dairy Sci. 88:2117-2129.

Benavides J.E. 1994. La Investigación en arboles Forrajeros. In: árboles y Arbustos Forrajeros en América Central. CATIE, Turrialba, CR, 1: 3-28.

Benavides, J.A., R.M. Ayala, M.R. Merino, 1996, Bellezas Naturales de Nuevo León, una guía de campo, Patronato Monterrey 400, CEMEX, Nuevo León p. 43-63.

Bergman, E. N. 1990. Energy contributions of volatile fatty acids from the gastrointestinal tract in various species. Physiology rev. 70: 567-590.

Beverly, C.A., Goff C.M., T. Forbes, D.A. 1997. "Toxic amines and alkaloids from Acacia berlandieri. Phytochemistry 46: 249-254.

Bird, A. R., R. D. Chandler, and A. W. Bell. 1981. Effects of exercise and plane of nutrition on nutrient utilization by the hind of the sheep. Aust. J. Biol. Sci. 34:541.

Blumenkrantz, N., and G. Asboe-Hansen. 1973. New method for quantitative determination of uronic acids. Analytical Biochemistry 54:484–489.

Bouderoua, K., J. Mourot, G. Selselet-Attou. 2009. The Effect of Green Oak Acorn (Quercus ilex) Based Diet on Growth Performance and Meat Fatty Acid Composition of Broilers. Asian Australas. J. Anim. Sci. 22, 843-848

Bourgaud, F., Gravot, A., Milesi, S. y Gontier, E. 2001. Production of plant secondary metabolites: a historical perspective. Plant Sci, 161: 839-851.

Bowers, J.E. 2007. Has climatic warming altered spring flowering date of Sonoran Desert shrubs? Southwestern Naturalist, 52: 347-355.

Bowers, J.E., Dimmitt, M.A. 1994. Flowering phenology of six woody plants in the northern Sonoran Desert. Bulletin of the Torrey Botanical Club, 121: 215-229.

Bozzo, Joseph A.; Beasom, Samuel L.; Fulbright, Timothy E. 1992. Vegetation responses to 2 brush management practices in south Texas. Journal of Range Management. 45(2): 170-175.

Brooker R.J., Widmaier E.P., Graham L.E. & Stiling P.D. 2008: Biology. McGraw-Hill. New York.

Brutsch, M.O. 1997. The beles or cactus pear (Opuntia ficus indica) in Tigray, Ethiopia. J. Prof. Assn. Cactus Devel., 2: 130-141.

Bryant, Fred C.; Demarais, Steve. 1991. Habitat management guidelines for whte-tailed deer in south and west Texas. In: Lutz, R. Scott; Wester, David B., editors. Research highlights--1991: Noxious brush and weed control; range and wildlife management. Volume 22. Lubbock, TX: Texas Tech University, College of Agricultural Sciences: 9-13.

Buxton D.R. & Redfearn D.D. 1997. Plant limitations to fibre digestion and utilisation. Journal of Nutrition 127: 814S-818S.

Caballero-Mellado, J. 1990. Potential use of Azospirillum in association with prickly pear cactus. Proceedings of First Annual Texas Prickly Pear Council. pp. 14-21.

Campbell, N. A. & Reece, J.B 2005. Biology (7th Edition). Benjamin Cummings Publishing.

Campbell, T. A. 1999. Antler development and nutritional influences of plant secondary compounds in mature white-tailed deer. M.S. Thesis, Texas A&M University-Kingsville, Texas, USA.

Campbell, T.A., and D.G. Hewitt. 2000. Effect of metabolic acidosis on white-tailed deer antler development. Physiological and Biochemical Zoology 73:781–789.

Carro M.D. 2001. La determinación de la sintesis de proteína microbiana en el rumen: Comparación entre marcadores microbianos (Revision). Invest. Agr. Prod. Sanid. Anim. 16:5-27.

Castillo Morales, N. E. 1997. Perfil nutricional de diez especies arbustivas de zonas semiáridas del sureste de Nuevo Leon. Tesis de licenciatura, Facultad de Ciencias Biológicas, UANL. Monterrey, N.L p. 10,32,33,34,35,37,38,39,40,41,43.

Chatterton, N.J., Goodin, R.J., Mckell, C.M., Braker, R.V. y Rible, J.R. 1971. Monthly variation in the chemical composition of desert saltbush. J. Range Manage. 24:37-40.

Chavez-Ramirez, F., X. Wang, K. Jones, D. Hewitt, and P. Felker. 1997. Ecological characterization of Opuntia clones in south Texas: Implications for wildlife herbivory and frugivory. J. Prof. Assn. Cactus Devel., 2: 9-19.

Clavero, T. 1996. Las leguminosas forrajeras arbóreas: Sus perspectiva para el trópico americano. In Clavero, T. (Ed) Leguminosas Forrajeras Arbóreas en la Agricultura Tropical. Centro de Transferencia de Tecnología en pastos y Forrajes Universidad del Zulia, pp. 1-10.

Clement, B.A., C.M. Goff, and T.D.A. Forbes. 1997. Toxic amines and alkaloids from Acacia berlandieri. Phytochemistry 46:249–254.

Cluff, L.K., Welch, B.L., Pederson, J.C., Brotherson, J.D., 1982. Concentration of sagebrush monoterpenes in the rumen ingesta of wild mule deer. J. Range. Manage. 35, 192–194.

Cooper, S.M., Owen-Smith, N., 1986. Effects of plant spinescence on large mammalian herbivores. Oecologia 68, 446–455.

Correll, S. D. & C. M. Johnston. 1970. Manual of the Vascular Plants of Texas. Texas Research Foundation. Renner, Texas, pp.1879.

COTECOCA. 1969. Coeficientes de Agostadero de la República Mexicana: Estados de Baja California, Sonora, Chihuahua, Zacatecas, Coahuila, Tamaulipas, Nuevo León Durango y San Luis Potosí. SAG, México, D.F.

Croteau, R. Kutchan, T.M., Lewis, N.G. 2000. Natural Products (Secondary Metabolites)". En: Buchanan, Gruissem, Jones (editores). Biochemistry and Molecular Biology of Plants. American Society of Plant Physiologists. Rockville, Maryland, Estados Unidos de América.

Davis, E. 1990. Deer management in the South Texas Plains. Austin, TX: Texas Parks and Wildlife, Federal Aid Report Serial 27. Contribution of Federal Aid (P-R) Project W 125-R.

Dawson, R. J., H. M. Armleder, and M. J. Waterhouse. 1990. Preference of mule deer for Douglas-fir foliage from different sized trees. Journal of Wildlife Management 54:378–382.

De Kock, G.C. 2001. The use of Opuntia as a fodder source in arid areas of southern Africa. p. 101–105. In C. Mondragon-Jacobo and S. Perez-Gonzalez (ed.) Cactus (Opuntia spp.) as forage. FAO plant production and protection paper 169, FAO, Rome, Italy.

De la Parra M.R. 1994. Estudio experimental sobre la producción vegetativa de Barreta (Helietta parvifolia) Gray (Benth), mediante las técnicas de estacas y cultivo de tejidos (in vitro). Tesis (profesional) Facultad de Ciencias Biológicas. UANL. Monterrey Nuevo León. Pp 1-7.

DeAngelis, D.L., Gross, L.J., 1992. Individual-based Models and Approaches in Ecology. Chapman & Hall, New York.

DeKock, G.C. 1980. Drought resistant fodder shrubs in South Africa. p. 399–408. In H.N. LeHouerou (ed.) Browse in Africa. The current state of knowledge. Papers presented at the International Symposium on Browse in Africa, Addis Ababa, International Livestock Center for Africa.

Delgiudice, G.D., L.D. Mech, and U.S. Seal. 1994. Nutritional restriction and acid base balance in white-tailed deer. Journal of Wildlife Diseases 30:247–253.

Devendra, C. 1993. Feed value of browse plants. Proceedings of VII World Conference on Animal Production, Edmonton, Alberta, Canada, pp. 119-136.

Díaz-Romeau, R. A. Hunter. 1978. Metodología del muestreo de suelos y tejidos de investigación en invernadero. CATIE(mimeo). Turrialba, Costa Rica.

Ditchkoff, S. S. and F. A. Servello. 1998. Litterfall: An overlooked food source for wintering white-tailed deer. Journal of Wildlife Management62:250–255.

Dominguez-Gomez, T.G, H. Gonzalez-Rodriguez, M. Guerrero-Cervantes, M.A. Cerrillo-Soto, A.S. Juarez-Reyes, M.S. Alvarado, and R.G. Ramirez-Lozano 2011. Polyethylene glycol influence on in vitro gas production parameters in four native forages consumed by White-tailed deer. J. Environ. Forest Sci. XII. Special Edition, pp. 21-32.

Dumont, A., J-P. Ouellet, M. Crête, and J. Huot. 2005. Winter foraging strategy of white-tailed deer at the northern limit of its range. Écoscience12:476–484.

Dupraz, C., 1999. Fodder trees and shrubs in Mediterranean areas: browsing for the future? In: Papanastasis, V., Frame, J., Nastis, A. (Eds.), Grasslands and Woody Plants in Europe. International Symposium, vol. 4. Thessaloniki, May 27–29, 1999. European Grassland Federation. Grassland Science in Europe, pp. 145–158.

Dziba, L.E., Scogings, P.F., Gordon, I.J., Raats, J.G., 2003. Effects of season and breed on browse species intake rates and diet selection by goats in the False Thornveld of the Eastern Cape, South Africa. Small Rumin. Res. 47, 17–30.

Ellis, J. L., F. Qiao, and J. P. Cant. 2006. Evaluation of net energy expenditures of dairy cows according to body weight changes over a full lactation. J. Dairy Sci. 89:1546-1557.

Esparza, Ch. 1980. Variación estacional de los atributos nutricionales de Atriplex canescens. Tesis de licenciatura, Escuela Superior de Agricultura y Zootecnia. Universidad Juárez del Estado de Durango. Gómez Palacio, Durango. México.

Esparza, H.J. y Ortiz, A. 1988. Fertilidad, prolificidad y porcentaje de abortos en dos períodos de ahijadero de tres hatos caprinos del Altiplano Potosino-Zacatecano. Centro regional de Estudios de Zonas Áridas y Semiáridas. Colegio de Posgraduados. Salinas S.L.P. México.

Espejel-Rodríguez, M.M.A., Santacruz-García, N. y Sánchez-Flores, M. 1999. El uso de los encinos en la región de La Malinche, Estado de Tlaxcala, México. Boletin de la Sociedad Botanica de México. 64:35-39.

Estrada, A.E., Jurado, E., 2005. Leguminosas del norte del estado de Nuevo León, México. Acta Bot. Mex. 73:1-18.

Etzenhouser, M.J., Owens, M.K., Spalinger, D.E. y Murden, S.B. 1998. Foraging behavior of browsing ruminants in a heterogeneous landscape. Landscape Ecology 13:55-64.

Everitt J.H., Drawe L.D. 1993. Texas Teach University Press, pp. 998.

Everitt, J.H.; Drawe, D.L.; Lonard, R.I. 2002. Trees, Shrubs and Cacti of South Texas. Texas Tech University Press, Lubbock, Texas, USA pp. 12-24.

Felker, P. 1995. Forage and fodder production and utilization. p. 144–154. In Inglese et al. (ed.) Agroecology, cultivation and uses of cactus pear. FAO Plant Production Paper 132. Rome, Italy.

Felker, P., S.C. Rodriguez, R.M. Casoliba, R. Filippini, D. Medina, and R. Zapata. 2005. Comparison of Opuntia ficus indica varieties of Mexican and Argentine origin for fruit yield and quality in Argentina. J. Arid Environ., 60: 405-422.

Fierro, L.C. 1991. Utilización de Atriplex canescens y su importancia en la dieta del ganado y su manejo. En: Memoria del taller sobre captación y aprovechamiento del agua con fines agropecuarios en zonas de escasa precipitación. Eds. H. Salinas, S. Flores, M. Martínez. INIFAP. SARH. p 255-270.

Flinn, Robert C.; Scifres, Charles J.; Archer, Steven R. 1992. Variation in basal sprouting in co-occurring shrubs: implications for stand dynamics. Journal of Vegetation Science. 30(1): 125-128.

Flores, C., and G. Aranda. 1997. Opuntia based ruminant feeding systems in Mexico. J. Prof. Assn. Cactus Devel., 2: 3-9.

Foley, W.J., and C. Mcarthur. 1994. The effects and costs of ingested allelochemicals in mammals: an ecological perspective. In: D. J. Chivers, and P. Langer [eds.]. The digestive system in mammals: Food, form and function. Cambridge, UK: Cambridge University Press, p. 370–391.

Foley, W.J., S. Mclean, and S.J. Cork. 1995. Consequences of biotransformation of plant secondary metabolites on acid-base metabolism in mammals-a final common pathway? Journal of Chemical Ecology 21:721–743.

Fonseca A.J.M., Dias-da-Silva A.A., Orskov E.R. 1998. *In sacco* degradation characteristics as predictors of digestibility and voluntary intake of roughages by mature ewes. Anim. Feed Sci. Tech., 72, 205–219.

Forbes, T.D.A., I.J. Pemberton, G.R. Smith, and C.M. Hensarling. 1995. Seasonal variation of two phenolic amines in Acacia berlandieri. Journal of Arid Environments 30:403–415.

Forbes, J. M. 1996. Integration of regularity signals controlling forage intake in ruminants. J. Anim. Sci. 74(12): 3029-35.

Foroughbackhch, R., RG. Ramirez, L.A. Hauad, N.E. Castillo-Morales, and J. Moya-Rodríguez. 1997. Seasonal dynamics of the leaf nutrient profile of 10 native shrubs of northeastern Mexico. Forest, Farm, and Community Tree Research Reports – Vol. (2): 8-12.

Foroughbackhch, R., Hernández, P.J.L., Alvarado, V.M.A., Céspedes, C.E., Rocha, E.A., Cárdenas, A.M.L., 2009. Leaf biomass determination on woody shrub species in semiarid zones. Agroforest. Syst. 77:181-192.

France, J., Theodorou, M.K., Lowman, R.S., Beever, D.E., 2000. Feed evaluation for animal production. In: M.K. Theodorou and J. France. (eds.) Feeding systems and feed evaluation models. CAB International, Wallingford, UK, pp 1-9.

Fulbright, T.E., Ortega, S.A. 2006. White-tailed Deer Habitat: Ecology and Management on Rangelands. Texas A&M University Press, pp 63-94.

Gabel, G, M. Marek and H. Martens. 1993. Influence of food deprivation on SCFA and electrolyte transport across sheep reticulo-rumen. Journal of Veterinary Medicine Series-A. 1993, 40: 5, 339-344; 3 ref.

Galizzi, F., P. Felker, and G. Gardiner. 2004. Correlations between soil and cladode nutrient concentrations and fruit yield and quality in Opuntia ficus indica in a traditional farm setting in Santiago del Estero, Argentina. J. Arid Environ., 59: 115–132.

Gallaher, Timothy, and Mark Merlin. 2010. Biology and impacts of Pacific Island invasive species. 6. Prosopis pallida and Prosopis juliflora (Algarroba, Mesquite, Kiawe)(Fabaceae). Pacific Science, 64: 489-526.

Ganskopp, D. 2001. Manipulating cattle distribution with salt and water in large arid-land pastures: a GPS/GIS assessment. Applied Animal Behavior Science 73:251–262.

Ganskopp, D., R. Cruz, D.E. Johnson. 2000. Least-effort pathways? A GIS analysis of livestock trails in rugged terrain. Applied Animal Behavior Science 68:179–190.

Getachew, G., E.J. DePeters, P.H. Robinson, and J.G. Fadel. 2005. Use of an in vitro rumen gas production technique to evaluate microbial fermentation of ruminant feeds and its impact on fermentation products. Anim. Feed Sci. Technol. 123/124:547-559.

Getachew, G., Makkar, H.P.S., Becker, K., 2000a. Effect of polyethylene glycol on in vitro degradability of nitrogen and microbial protein synthesis from tannin-rich browse and herbaceous legumes. Brit. J. Nutr. 84, 73–83.

Getachew, G., Makkar, H.P.S., Becker, K., 2000b. Effect of different amounts and method of application of polyethylene glycol on efficiency of microbial protein synthesis in an in vitro system containing tannin rich browses. In: Proceedings of the EAAP Satellite Symposium, Gas Production: Fermentation Kinetics for Feed Evaluation and to Assess Microbial Activity. Brit. Soc. Anim. Sci., and Wageningen University, PUDOC, Wageningen, The Netherlands, p. 93.

Getachew, G., Makkar, H.P.S., Becker, K., 2001. Method of polyethylene glycol application to tannin containing browses to improve microbial fermentation and efficiency of microbial protein synthesis from tannin-containing browses. Anim. Feed Sci. Technol. 92, 51–57.

Gilman, E.F. 1999. Caesalpinia mexicana Mexican Caesalpinia. IFAS Extension. University of Florida.

Goering, H K., and P.J. Van Soest. 1970. Forage analyses (apparatus, reagents, procedures and some applications). USDA Agricultural Handbook 379. 20 p.

González, P.I. y Ortega, S.J.L. 1994. Determinación de los coeficientes de digestibilidad aparente de forraje de dos especies de Atriplex: A. canescens y A. acanthocarpa en ovinos. Tesis. Universidad Autónoma Chapingo. Unidad Regional Universitaria de Zonas Áridas, Bermejillo, Durango.

González, R.H., Ramírez, L.R.G., Cantú, S.I., Gómez, M.M.V., Uvalle, S.J.I., 2010. Composición y estructura de la vegetación en tres municipios del estado de Nuevo León. Polibotánica 29:91-106.

González, R.H., Ramírez, L.R.G., Cantú, S.I., Gómez, M.M.V., Uvalle, S.J.I., 2010. Composición y estructura de la vegetación en tres municipios del estado de Nuevo León. Polibotánica 29:91-106.

González-Gómez, J.C., Ayala-Burgos, A. y Gutiérrez-Vázquez, E. 2006. Determinación de fenoles totales y taninos condensados en especies arbóreas con potencial forrajero de la Región de Tierra Caliente Michoacán, México. Livestock Research for Rural Development. Volume 18, Article No. 152. Obtenido en Febrero 10, 2010, de: http://www.lrrd.org/lrrd18/11/guti18152.htm.

Gonzalez-Rodriguez, H., I. Cantu-Silva, M.V. Gomez-Meza, and R.G. Ramirez-Lozano 2004. Plant water relations of thornscrub shrub species, northeastern Mexico. J. Arid Environ. 58:483-503.

Gonzalez-Rodriguez, H., T.G. Domínguez-Gomez, I. Cantú-Silva, M.V. Gómez-Meza, R.G. Ramirez-Lozano, M. Pando-Moreno, and C.J. Fernandez. 2011. Litterfall deposition and leaf litter nutrient return in different locations at Northeastern Mexico. J. Plant Ecol. 212:1747-1757.

Greene, L.W., Pinchak, W.E., Heitschmidt, R.K. 1987. Seasonal dynamics of mineral in forages at the Texas Experimental Ranch. Journal of Range Management, 40: 502-510.

Grings, E.E., M.R. Haferkamp, R.K. Heitschmidt, and M.G. Karl. 1996. Mineral dynamics in forages of the Northern Great Plains. J. Range Manage. 49: 234-240.

Guerrero, M, A.S. Juarez, R.G. Ramirez, R. Montoya, O. La O. M. Murillo, and S.M.A. Cerrillo. 2010. Chemical composition and protein degradability of native forages of the semiarid region of Northern Mexico. Cuban J. Agric. Sci. 44:143- 149.

Guerrero, M., Cerrillo, S.M.A., Ramírez, R.G., Salem, A.Z.M., González, H., Juárez, R.A.S., 2012. Influence of polyethylene glycol on in vitro gas production profiles and microbial protein synthesis of some shrub species. Anim. Feed Sci. Tech. 176:32-39.

Guerrero-Cervantes, M., Ramirez, R.G., Cerrillo-Soto, M.A. Montoya-Escalante, R., Nevárez-Carrasco, G., Juárez-Reyes, A.S. 2009. Chemical composition and rumen digestion of dry matter and crude protein of native forages. Journal of Animal and Veterinary Advances, 8(8): 408-412.

Guerrero-Cervantes, M., Ramírez, R.G., Cerrillo-Soto, M.A., Montoya-Escalante, R., Nevárez-Carrasco, G., Juárez-Reyes, A.S., 2009. Dry matter digestion of native forages consumed by range goats in North Mexico. J. Anim. Vet. Adv. 8(3):408-412.

Guglielmelli, A., Calabrò, S., Primi, R., Carone, F., Cutrignelli, M.I., Tudisco, R., Piccolo, G., Ronchi, B., Danieli, P.P., 2011. In vitro fermentation patterns and methane production of sainfoin (Onobrychis viciifolia Scop.) hay with different condensed tannin contents. J. Brit. Grassland Soc. 1-13.

Guglielmelli, A., Calabrò, S., Primi, R., Carone, F., Cutrignelli, M.I., Tudisco, R., Piccolo, G., Ronchi, B., Danieli, P.P., 2011. In vitro fermentation patterns and methane production of sainfoin (Onobrychis viciifolia Scop.) hay with different condensed tannin contents. J. Brit. Grassland Soc. 1-13.

Gutteridge, R.C. y M.H. Shelton. 1994. The role of forage tree legumes in cropping and grazing systems. En: Forage tree legumes in tropical agriculture. CAB International. Wallingford, UK. P 3-11.

Gutteridge, R.C. y M.H. Shelton. 1994. The role of forage tree legumes in cropping and grazing systems. En: Forage tree legumes in tropical agriculture. CAB International. Wallingford, UK. P 3-11.

Hanley, T.A., C.T. Robbins, A.E. Hagerman, and C. Mcarthur. 1992. Predicting digestible protein and digestible dry matter in tannin-containing forages consumed by ruminants. Ecology 73:537–541.

Hanley, M. E. et al. 2007. Plant structural traits and their role in anti-herbivore defence. Perspectives in Plant Ecology Evolution and Systematics. 8: 157-178.

Harden, M. L., and Reza Zolfaghari.1988. Nutritive composition of green and ripe pods of honey mesquite (Prosopis glandulosa, Fabaceae). Economic Botany, 42: 522-532.

Harris, N.R., D.E. Johnson, M. R. George, N.K. McDougald. 2002. The effect of topography, vegetation, and weather on cattle distribution at the San Joaquin Experimental Range, California. USDA Forest Service General 58(2) March 2005 117 Technical Report PSW-GTR-184. p 53–63. Available from: USDA Forest Service, Albany, CA.

Hassan, S.W., Umar, R.A., Ebbo, A.A., Matazu, I.K. 2005. Phytochemical, antibacterial and toxicity studies of Parkinsonia aculeata L. (Fabaceae). Nigerian Journal Biochemistry Molecular Biology, 20(1): 89-97.

Hellgren, E.C., and W.J. Pitts. 1997. Sodium economy in white-tailed deer (Odocoileus virginianus). Physiological Zoology 70:547–555.

Hernandez G, R. 1997. Análisis estrictural e importancia economica de Helietta parvifolia Gray (Benth), en dos zonas ecológicas en el estado de Nuevo León. Tesis (profesional) Facultad de Ciencias Biológicas, UANL. San Nicolás de los Garza, Nuevo León, México. Pp 6-16, 35.

Hernández, G.M.H. 1985. Características ecológicas, fisiológicas y forrajeras del género Atriplex. Tesis. Ing. Agrónomo. Universidad Autónoma Chapingo.

Hernández, R. S. 1981. Especies arbóreas forestales susceptibles de aprovecharse como forraje. Ciencia Forestal, INIF, 6: 31-39.

Hewitt, D.G., and R.L. Kirkpatrick. 1997. Ruffed grouse consumption and detoxification of evergreen leaves. Journal of Wildlife Management 61:129–139.

Hoffman, P.C., Lundberg, K.M., Bauman, L.M., Shaver, D.R., Contreras, G.F.E., 2007. The effect of maturity on NDF (neutral detergent fibre) digestibility. Focus on forage. University of Wisconsin ed., Madison, WI, USA.

Hofmann, R. R. 1988. Anatomy of the gastro-intestinal tract. In: D. C. Church (Ed) The Ruminant Animal: Digestive Physiology and Nutrition. Prentice Hall, Englewood Cliffs, NJ, pp. 14-43.

Holechek, J.L., A.V. Munshikpu, L. Saiwana, G. Nuñez-Hernández, R. Valdez, J.D. Wallace and M. Cardenas. 1990. Influences of six shrubs diets varying in phenol content on intake and nitrogen retention by goats. Tropical Grasslands, 24: 93-98.

Holter, J.B., S.H. Smith, and H.H. Hayes. 1979. Protein requirements of yearling white-tailed deer. Journal of Wildlife Management 43:872–879.

Hoste, H., Jackson, F., Athanasiadou, S., Thamsborg, S.M., Hoskin, S.O. 2006. The effects of tannin-rich plants on parasitic nematodes in ruminants. Trends Parasitol. 22: 253-261.

Hughes, H. G. 1982. Estimated energy, protein, and phosphorus balances of a south Texas white-tailed deer population [Ph.D. thesis]. College Station, TX: Texas A&M University.

Hughes. C.E. and Styles, B.T. 1984. Exploration and seed collection of multipurpose dry zones trees in Central America. International Tree Crops Journal, 3: 1-31.

Ibarra, F.A., Garza, H.M. y de Luna, R. 1977. Establecimiento de costilla de vaca (Atriplex canesces Pursh Nutt) en forma directa, bajo estructuras de poceo en condiciones áridas. UAAAN. Saltillo, Coah. Monog. Tec. Cient. 5:49-123

Irish, M. 2008. Trees and Shrubs for the Southwest: Woody Plants for Arid Gardens. Timber Press. pp. 205–206.

Jančik F., Homolka P., Čermak B., Lad F. (2008): Determination of indigestible neutral detergent fiber contents of grasses and its prediction from chemical composition. Czech Journal of Animal Science, 53: 128–135.

Jean-Blain, C. 1998. Aspectes nutritionnels et toxicologiques des tannins. Rev. Méd. Vét. 149: 911-920.

Jenks, J. A. and D. M J. Leslie. 1989. Digesta retention of winter diets in white-tailed deer Odocoileus virginianus fawns in Maine, USA. Canadian Journal of Zoology 67:1500–1504.

Juárez-Reyes, A.S. Nevarez-Carrasco, G., Cerrillo-Soto, M.A., Murillo-Ortiz, M., Luginbuhl, J.M., Bernal-Barragan, H., Ramírez, R.G. 2008. Dietary Chemical Composition, Plasma Metabolites and Hormones in Range Goats. Journal of Applied Animal research, Res. 34(1): 81-86.

Kahn, S. R [ed.]. 1995. Calcium oxalate in biological systems., Boca Raton, FL: CRC Press.

Kamalak A, Canbolat O, Gurbuz Y, Ozay O, Ozkan C O and Sakarya M. 2004. Chemical composition and in vitro gas production characteristics of several tannin containing tree leaves. Livestock Research for Rural Development, Vol. 16, Art.#44. ht tp://www.cipav.org.co/lrrd/lrrd16/6/kama16044.htm.

Kazemi, M., A.M. Tahmasbi, A.A. Naserian, R. Valizadeh, and M.M. Moheghi. 2012. Potential nutritive value of some forage species used as ruminants feed in Iran. African J. Biotechnol. 11:1210-1217.

Kendall, P. E. ; McLeay-L. M. 1996. Excitatory effects of volatile fatty acids on the un vitro motility of the rumen of sheep. Res. Vet. Sci. 1996. July 6(1) 1-6.

Kessler, J.J. 1990. Atriplex forage as a dry season supplementation feed for sheep in the montane plains of the Yemen Arab Republic, Journal of Arid Enviroments. 19:225-234.

Khennouf, S., S. Amira, L. Arrar y A. Baghiani. 2010. Effect of Some Phenolic Compounds y Quercus Tannins on Lipid Peroxidation. World Appl. Sci. J., 8 (9): 1144-1149.

Kilic, U., M. Boga y I. Guven. 2010. Chemical Composition and Nutritive Value of Oak (Quercus robur) Nut and Leaves. J. Applied Animal Research, 38: 101-105.

Klemm, D., Heublein B., Fink, H. & Bohn, A. 2005: Cellulose: fascinating biopolymer and sustainable raw material. ChemInform 36: 3358-3393.

Kohn, R. A. 1994. Equilibrium concentrations of volatile fatty acids in the rumen. National Conf. On forage Quality, Evaluation and Utilization. University of Nebraska-Lincoln. P. 51.

Kohn, R. A. 1994. Equilibrium concentrations of volatile fatty acids in the rumen. National Conf. On forage Quality, Evaluation and Utilization. University of Nebraska-Lincoln. P. 51.

Krause, D. O., S. E. Denman, R. I. Mackie, M. Morrison, A. L. Rae, G. T. Attwood, and C. S. McSweeney. 2003. Opportunities to improve fiber degradation in the rumen: Microbiology, ecology, and genomics. FEMS Microbiol. Rev. 27: 663-693.

Krehbiel, C. R., C. A. Bandyk, M. J. Hersom, and M. E. Branine. 2008. Alpharma beef cattle nutrition symposium: Manipulation of nutrient synchrony. J. Anim. Sci. 86: E285-E286.

Kuehl, R. O. 1994. Statistical principles of research design and analysis. Belmont, CA: Duxbury Press. 686 p.

Kumar, R., Vaithiyanathan, S., 1990. Occurrence nutritional significance and effect on animal productivity of tannins in tree leaves. Anim. Feed Sci. Technol. 30, 21–38.

Laca, E.O., Shipley, L.A., Reid, E.D. 2001. Structural anti-quality characteristics of range and pasture plants. J. Range Manage. 54: 413–419.

Lambers, H. et al. 1998. Plant physiological ecology. New York, NY: Springer Science.

Landou, S.Y., Perevolotsky, A., Kababya, D., Silanikove, N., Nitzan, R., Baram, H., Provenza, F.D., 2002. Polyethylene-glycol increases the intake of tannin-rich Mediterranean browse by ranging goats. J. Range Manage. 55: 598–603.

Launchbaugh, K.L., Provenza, F.D., Pfister, J.A., 2001. Herbivore response to anti-quality factors in forages. J. Range Manage. 54: 431–440.

Ledezma Menxueiro, A.R. 2001. Fitosociología de Larrea tridentata en el matorral micrófilo en los municipios de mina, Nuevo León y Castaños, Coahuila, México. Tesis de Maestría, Facultad de Ciencias Biológicas, UANL.

Leonard E.N. 2004. Etymological Dictionary of Succulent Plant Names. Birkhäuser. p. 11.

Lerma Hernández, Alvaro. 1988. Estudio de la variación estacional en el contenido de nutrientes y digestibilidad de Cenchrus ciliaris L., Acacia rigidula Benth. y Atriplex nummularia Lindl. en la región semiárida del noreste de México.

Lincoln, T. y Zeiger, E. 2006. Secondary Metabolites and Plant Defense. Plant Physiology, Fourth Edition. Sinauer Associates, Inc. EUA.

Lonard, Robert I.; Judd, Frank W. 1991. Comparison of the effects of the severe freezes of 1983 and 1989 on native woody plants in the lower Rio Grande Valley, Texas. Southwestern Naturalist. 36(2): 213-217.

Lopez-Coba, E., C.A. Sandoval-Castro and R.C. Montes-Perez, 2007. Intake and Digestibility of Tree Fodders by White Tailed Deer (Odocoileus virginianus yucatanensis). Journal of Animal and Veterinary Advances, 6: 39-41.

Lowry, B.J., Petherman, J.R., Tangenjaja, T. 1992. Plants fed to village ruminants in Indonesia. ACIAR, Technical Report No. 22, Canberra, p. 60.

Lozoya, Xavier. 1976. Estado Actual del Conocimiento en las Plantas Medicinales Mexicanas. 1ª.edición. Instituto Mexicano para el estudio de plantas naturales medicinales. A.C. p.p. 144

Luchow, M. 1993. Monograph of Desmanthus (Leguminosae-Mimosoideae). Sys. Botanical Monographs 38:113-117.

Madrigal, X. 1997. Características generales de la vegetación del estado de Durango, México. Ciencia Forestal, 27: 30-58.

Magallanes C, E. 1985. Evaluación de los efectos fisiológicos y anatómicos causados por diferentes estractos de Helietta parvifolia Gray (Benth), especie aleopática en algunas especies de plantas cultivadas. Tesis (profesional) Facultad de Ciencias Biológicas, UANL. Monterrey Nuevo León, México. Pp 10-14.

Makkar, H.P.S. 2003. Effects and fate of tannins in ruminant animals, adaptation to tannins, and strategies to overcome detrimental effect of feeding tannin-rich feeds. Small Rumin. Res. 49:241-256.

Makkar, H.P.S. 2006. Chemical and biological assays for quantification of major plant secondary metabolites. BSAS Publication 34. The assessment of intake, digestibility and the roles of secondary compounds. Edited by C.A. Sandoval-Castro, F.D.DeB.D. Hovell, J.F.J. Torres-Acosta and A. Ayala-Burgos. Nottingham University Press. pp. 235-249.

Maldonado, L.J. 1979. Caracterización y usos de los recursos naturales de las zonas áridas. Ciencia Forestal, 4: 56-64.

Mangione, A.M., Dearing, M.D., Karasov, W.H. 2000. Interpopulation differences in tolerance to creosote bush resin in desert woodrats (Neotomalepida). Ecology, 81(8): 2067-2076.

Marini, J. C., and M. E. Van Amburgh. 2003. Nitrogen metabolism and recycling in Holstein heifers. J. Anim. Sci. 81: 545-552.

Marini, J. C., J. D. Klein, J. M. Sands, and M. E. Van Amburgh. 2004. Effect of nitrogen intake on nitrogen recycling and urea transporter abundance in lambs. J. Anim. Sci. 82: 1157-1164.

Martin, J.S., and M.M. Martin. 1982. Tannin assays in ecological studies: lack of correlation between phenolics, proanthocyanidins and protein-precipitating constituents on mature foliage of six oak species. Oecologia 54: 205–211.

Martínez Maximino. Catalogo de nombres vulgares y científicos de plantas mexicanas. Edit. Fondo de cultura económica. 1era edición. P: 1078.

Martinez, M. 1994. Plantas Mexicanas. 1a ed. Fondo de Cultura Economica. Mexico, pp. 205-230.

Martínez, S.J., Pedraza, R.M. y García, Y. 2001. Influencia del método de secado del follaje y el solvente de extracción en la cuantificación de polifenoles extractables totales. Pastos y Forrajes 24: 353-356.

Massey, F. P. & Hartley, S. E. (2009). Physical defences wear you down: progressive and irreversible impacts of silica on insect herbivores. Journal of Animal Ecology 78, 281-291.

McDonald, I.1981. A revised model for the estimation of protein degradability in the rumen. J. Agric. Sci., Camb., 96 : 251 - 252.

McDowell, L.R. and G. Valle. 2000. Macro minerals in forages. In: Forage Evaluation in Ruminant Nutrition (Eds.): D.I. Givens, E. Owen, R.F.E. Oxford and H.M. Omed. CAB International, pp. 136-141.

McMullen, C. K. 1999. Flowering plants of the Galápagos. Comstock Pub. Assoc., Ithaca, N.Y. p. 186.

McSweeney, C.S., Dalrymple, E.P., Gobius, K.S., Kennedy, P.M., Krause, D.O., Lowry, J.B., Mackie, R.I. and Xue, G.P. 1998. The application of rumen biotechnology to improve the nutritive value of fibrous feedstuffs: pre and post ingestion. In: Proceedings of 8th World Conference on Animal Production. Seoul National University, Seoul Korea, pp. 392-421.

Melaku, S., T. Aregawi, and L. Nigatu (2010). Chemical composition, in vitro dry matter digestibility and in sacco degradability of selected browse species used as animal feeds under semi-arid conditions in Northern Ethiopia. Agrofor. Syst. 80:173-184.

Mellado, M. 1990. Producción de caprinos en pastoreo. UAAAN, Buenavista, Coahuila. México. 3. Arbiza, A.S.I. 1986. Nutrición y alimentación. En: producción de caprinos. Ed. AGT Editor S.A. México. pp 310-312.

Menke, K.H., and H. Steingass (1988). Estimation of the energetic feed value obtained from chemical analysis and gas production using rumen fluid. Anim. Res. Develop. 28:7-55.

Mercado-Santos, A.C. 1999. Dinámica estacional del valor nutritivo y degradabilidad ruminal de la material seca y proteína cruda de 15 plantas arbustivas del noreste de Nuevo León. Tesis de Licenciatura, Facultad de Veterinaria. Universidad Autónoma de Nuevo León.

Mertens, D.R. 2003. Challenges in measuring insoluble dietary fiber. J. Anim. Sci. 81:3233-3240.

Mertens, D.R. 1993. Kinetics of cell wall digestion and passage in ruminants. In: Jung, H. G., Buxton, D.R., Hatfield, R.D. and Ralph

J. (eds.). Forage Cell Wall Structure and Digestibility. ASA-CSSA-SSSA, Madison, WI., pp. 535-570.

MieHinen, H. and Huhtanen, P. 1996. Effects of the ratio of ruminal propionate to butyrate of milk yield and blood metabolites in dairy cows. J-Dairy-Sci, 1996 May; 79 (5): 851-61.

Mielke, J. 1993. Native Plants for Southwestern Landscapes. University of Texas Press. p. 158.

Minerals in Animal and Human Nutrition. 2nd edition, Elsevier, The Netherlands, pp. 256-315.

Minson, 1990. Forage in ruminant Nutrition. Academic Press, San Diego, p. 483.

Minson, D.L. 1990. Forage in Ruminant Nutrition. Academic Press, San Diego, pp. 162-168.

Mireles, R. E. A. 1997. Variación estacional de la concentración de ácidos grasos volátiles del forraje de 11 zacates nativos y cultivados en el noreste de México. Tesis de licenciatura. Facultad de Medicina Veterinaria y Zootecnia, UANL. Monterrey, Nuevo León.

Miyaki, M. and K. Kaji. 2004. Summer forage biomass and the importance of litterfall for a high-density sika deer population. Ecological Research19:405–409.

Montgomery, D.C. 2004. Experimental Designs. Second ed. Limusa-Wiley. DF, México, pp. 79-81.

Morrison, S. F., G. J. Forbes, and S. J. Young. 2002. Browse occurrence, biomass, and use by white-tailed deer in a northern New Brunswick deer yard. Canadian Journal of Forest Research 32:1518–1524.

Mould, E.D., and C.T. Robbins. 1981. Evaluation of detergent analysis in estimating nutritional value of browse. Journal of Wildlife Management 45: 937–947.

Moya Rodriguez, J.G. 1997. Dinamica estacional de la degradabilidad ruminal de los nutrientes contenidos en hojas de diez especies arbustivas nativas del noreste de México. Tesis de Maestria, Facultad de Ciencias Biológicas, UANL, Monterrey N.L.

Moya, Rdz, José guadalupe.2002. Variación estacional del perfil nutritivo y digestabilidad in situ de la amteria seca, proteína cruda y fibra detergente neutro, del follaje de 8 especies arbustivas del noreste de México. Tesisis de Doctorado. Facultad de Ciencias Biológicas. Universidad Autónoma de Nuevo León.

Moya-Rodríguez, J. G., Foroughbachkch, R., Ramirez, R.G. 2002. Variación estacional y digestibilidad in situ de la materia seca, de las hojas de arbustivas del noreste de México. International Journal of Experimental Botany, FYTON, 66(2): 121-127.

Mrvos, R. et al. (1991). Philodendron/Dieffenbachia ingestions: Are they a problem? Clinical Toxicology 29: 485-491.

Nantoume, H., Forbes, T. D.A., Hensarling, C.M. Sieckenius, S.S. 2001. Nutritive value and palatability of guajillo (Acacia berlandieri) as a component of goat diets. Small Ruminant Research, 40: 139-148.

National Research Council. 2001. Nutrient Requirements of Dairy Cattle. 7th rev. ed. ed. Natl. Acad. Press, Washington DC.

Nefzaoui, A., Ben Salem, H. and Ben Salem, L. 1996. Nitrogen supplementation of cactus based diets fed to Barbarine yearlings. In: « Native and Exotic Fodder Shrubs in Arid and Semi-Arid Zones, Regional Training Workshop, Tunisia, 27 October - 2 November, 1996.

Nelson, C.J., L.E. Moser. 1994. Plant factors affecting forage quality. In Forage quality, evaluation, and utilization. Editor George C. Fahey, Jr. (Univ of Nebraska, Lincoln, USA) 115-154.

Nelson, C.J., L.E. Moser. 1994. Plant factors affecting forage quality. In Forage quality, evaluation, and utilization. Editor George C. Fahey, Jr. (Univ of Nebraska, Lincoln, USA) 115-154.

Niembro R., A. 1986. Árboles y arbustos útiles de México. Ed. Limusa. México, D. F. 206 pp.

Nocek, J.E., and Tamminga, S. 1991. Site of digestion of starch in the gastrointestinal tract of dairy cows and its effect on mild yield and composition. J. Dairy Sci. 74: 3598-3629.

Nokes, J. 1986, How to grow native plants of Texas and the Southwest, Gulf Publishing Company, United States of America, pp. 12, 42, 228-230.

Nokes, J. 1986. How to grow native plants of Texas and the Southwest. Texas Monthly Press, Inc. Austin. 404 pp. McDowell L.R. 2003.

Nokes, J. 2001. How to Grow Native Plants of Texas and the Southwest (2 ed.). University of Texas Press. pp. 151-152.

Norton, B.W. and Poppi, D.P. 1994. Composition and Nutritional Attributes of Pasture Legumes. In: Tropical Legumes in Animal Nutrition. D´Mello, J. P. F. and Devendra, C. (eds.). Cab International, Wallingford, Reino Unido. pp. 23 – 48.

Norton, B.W., Poppi, D.P. 1995. Composition and nutritional attributes of pasture legumes. In: D'Mello, J.P.F., Devendra, C (eds.) Tropical Legumes in Animal Nutrition. CAB, International, Wallingford, pp. 23-48.

Nousiainen J., Ahvenjärvi S., Rinne M., Hellämäki M., Huhtanen P. 2004. Prediction of indigestible cell wall fraction of grass silage by near infrared reflectance spectroscopy. Anim. Feed Sci. Technol., 115, 295–311.

Nousiainen J., Rinne M., Hellämäki M., Huhtanen P. 2003. Prediction of the digestibility of the primary growth and regrowth grass silages from chemical composition, pepsin-cellulase solubility and indigestible cell wall content. Anim. Feed Sci. Technol., 110, 61–74.

NRC (National Research Council). 2007. Nutrient Requirements of Small Ruminants. Sheep, Goats, Cervids and New World Camelids. Washington DC., pp. 109-136.

NRC, 2007. National Research Council. Nutrient Requirement of Small Ruminant. Sheep, goats, cervids, and new world camelids. National Academy of Press. Washington, DC, pp. 166-178.

Núñez, J.F. y Simplicio A. 1980. Influencia de estacao de monta no nascimento de cabritos. Centro Nacional de Pesquisa de caprinos. Empresa Brasilira de Pesquisa Agropecuaria Brasil. Comunicado Técnico No.2, p 5.

O'Reilly, G. 2002. Tannin wars. Department of Business, Industry & Resource Develop.

Ørskov, E.R., I. McDonald. 1979. The estimation of protein degradability in the rumen from incubation measurements weighed according to rate of passage. J. Agri. Sci. (Cambridge), 92: 499-503

Ortega S., J.A., E.A. González V., J.M. Avila C. y R. Guarneros A. 1996. Manejo y utilización de Desmanthus para la alimentación de bovinos. XIV Día del Ganadero. INIFAP-CEAL- SAGAR. Public. Esp. No 3. Aldama, Tamps., México. P15-18.

Ortiz, R, D. 1992. Efecto de aceite esencial y agua de arrastre Helietta parvifolia en la inhibición de hongos postcosecha, que afecta frutos de tomate bajo condiciones de almacen. Tesis (profesional) Facultad de Ciencias Biológicas, UANL. Monterrey Nuevo León, México. Pp 17-18.

Osman, E., Muir, J., and Elgersma, A. 2002. Effect of Rhizobium Inoculation and Phosphorus Application on Native Texas Legumes Grown in Local Soil. J. Plant Nut. 25: 75-92.

Ouédraogo-Koné, S., Kaboré-Zoungrana, C.Y., Ledin, I. 2006. Behaviour of goats, sheep and cattle on natural pasture in the sub-humid zone of West Africa. Livestock Science,105(1): 224-252.

Papanastasis, V.P., Nefzaoui, A., 2000. Role of woody forage plants and cactus in livestock feeding in arid andsemi-arid Mediterranean areas. In: Guessous, E., Rihani, N., Alham, A. (Eds.), Livestock Production and Climatic Uncertainty in the Mediterranean EEAP. Wageningen Press, Wageningen, the Netherlands, No. 94,pp. 55-62.

Papanastasis, V.P., Yiakoulaki, M.D., Decandia, M., Dini-Papanastasi, O. 2008. Integrating woody species into livestock feeding in the Mediterranean areas of Europe. Animal Feed Science and Technology, 140(1-2): 1-17.

Parker, K.L., M.P. Gillingham, T. A. Hanley, and C. T. Robbins. 1999. Energy and protein balance of free-ranging black-tailed deer in a natural forest environment. Wildlife Monographs 143:1–48.

Parrotta, J.A. 1992. Acacia farnesiana (L.) Wild. aroma, huisache. SO-ITF-SM_49. New Orleans, LA: U.S. Department of Agriculture, Forest Service, southern Forest Experiment Station, p. 6.

Patra, A.K. 2009. Meta-analysis on effects of supplementing low-quality roughages with foliages from browses and tree fodders on intake and growth in sheep. Livestock Science, 121(2): 239-249.

Patra, A.K. 2010. Effects of supplementing low-quality roughages with tree foliages on digestibility, nitrogen utilization and rumen characteristics in sheep: a meta-analysis. Journal of Animal Physiology and Animal Nutrition, 94(2): 338–353.

Pekins, P.J., K.S. Smith, and W.W. Mautz. 1998. The energy cost of gestation in white-tailed deer. Canadian Journal of Zoology 76:1091–1097. REMINGTON, T. E. 1990. Food selection and nutritional ecology of blue grouse during winter [Ph.D. thesis]. Madison, WI: University of Wisconsin.

Pereira Filho J M, Vieira E L, Kamalak A, Silva A M A, Cezar M F e Beelen P M G 2005:Correlação entre o teor de tanino e a degradabilidade ruminal da matéria seca e proteína bruta do feno de jurema-preta (Mimosa tenuiflora Wild) tratada com hidróxido de sódio. Livestock

Research for Rural Development.Volume 17, Art. #91. http://www. cipav.org.co/lrrd/lrrd17/8/pere17091.htm.

Peter Bigfoot. 2011. Chaparral. Peter Bigfoot's Useful Wild Western Plants.

Pichersky, E. y Gang, D.R. 2000. Genetics and biochemistry of secondary metabolites in plants: an evolutionary perspective. Trends in Plant Sci. 5: 439-445.

Pieper R. and G. Donart. 1981. Fourwing saltbush response to short rest periods. In: Rangeland improvements for New México. New Mexico State Uni. Agricultural Exp. Station, Special report. p 41.

Potvin, F. and B. Boots. 2004. Winter habitat selection by white-tailed deer on Anticosti Island 2: Relationship between deer density from an aerial survey and the proportion of balsam fir forest on vegetation maps. Canadian Journal of Zoology 82:671–676.

Potvin, F., B. Boots, and A. Dempster. 2003. Comparison among three approaches to evaluate winter habitat selection by white-tailed deer on Anticosti Island using occurrences from an aerial survey and forest vegetation maps. Canadian Journal of Zoology 81:1662–1670.

Potvin, F., P. Beaupré, and G. Laprise. 2003. The eradication of balsam fir stands by white-tailed deer on Anticosti Island, Québec: A 150-year process. Écoscience 10:487–495.

Powell, A.M. 1998. Trees and Shrubs of the Trans-Pecos and Adjacent Areas. University of Texas Press. pp. 203–204.

Provenza, F.D., J.J. Villalba, L.E. Dziba, S.B. Atwood, R.E. Banner. 2003. Linking herbivore experience, varied diets, and plant biochemical diversity. Small Ruminant Research 49:257–274

Provenza, F.D. 2006. Behavioral mechanisms influencing use of plants with secondary metabolites by herbivores. BSAS Publication 34. The assessment of intake, digestibility and the roles of secondary compounds. Edited by C.A. Sandoval-Castro, F.D.DeB.D. Hovell, J.F.J. Torres-Acosta and A. Ayala-Burgos. Nottingham University Press. pp. 183-195.

Puig, Henri. 1991. Vegetación de la Huasteca (México). Estudio filogeográfico y ecológico. 121-149

Quiñones, V.J.J. 1987. Evaluación indirecta de la biomasa de Atriplex canescens en el noreste del estado de Durango. Tesis de maestría. Universidad Autónoma de Chihuahua, Facultad de Zootecnia. Chihuahua, Chih.

Quiñones, V.J.J. 1991. Carga de Atriplex y consideraciones metodológicas de evaluación. En: Memoria del taller sobre captación y aprovechamiento del agua con fines agropecuarios en zonas de escasa precipitación. Eds. H. Salinas, S. Flores, M. Martínez. INIFAP. SARH. p 255- 270.

Quiñones, V.J.J., Valencia, C.M., Sánchez, T. y Montañez, R. 1986. Variables que influyen sobre la producción de leche de caprinos en pastoreo de malezas y esquilmos en la Comarca Lagunera. Univ. Autónoma Agraria Antonio Narro. Saltillo, Coah. II Reunión Nacional sobre Caprinocultura. 12.

Ramirez Martinez, Hector Javier. 1991. Efecto de tres medios de propagacion y ocho metodos de preacondicionamiento en la germinacion de semillas de coma (Bumelia celastrina H.B.K.) bajo condiciones de invernadero en Marin, N.L. p. 1, 2, 4, 5.

Ramírez R. G., L. A. Hauad, R. Foroughbakhch and L. A. Pérez-López. 1997. Seasonal concentrations of in vitro volatile fatty acids in leaves of 10 native shrubs of northeastern of México. Forest, Farm and Community Tree Research reports - Vol. 2 (1997).

Ramirez RG, Mireles E, Huerta JM, Aranda J. 1995. Food habits of range sheep grazing a buffelgrass pasture. Small Ruminant Research, 17 :129-136.

Ramirez RG, Quintanilla-Gonzalez JB, Aranda J. 1997. Diets of white-tailed deer in northeastern Mexico., Small Ruminant Research, 25:142-149.

Ramirez RG. 1996. Feed value of browse. Proceedings of V International Conference on Goats. International Academic Publishers, Beijing China, 510.

Ramirez RG. 1996. Feed value of browse. Proceedings of V International Conference on Goats. International Academic Publishers, Beijing China, 510.

Ramírez, L.R.G., Háuad, L.A., Foroughbackhch, R. and Moya-Rodríguez, J. 1998b. Extent and rate of digestion of the dry matter in leaves of 10 native shrubs form northeastern México. International Journal of Experimental Botany. 62:175-180.

Ramírez, L.R.G., Ledezma-Torres R.A., Peña-Hernández A.F. and Moreno-Villanueva R. 1998a. Dry matter and protein degradation of foliage from Medicago sativa, Acacia greggii and Prosopis glandulosa by sheep. International Journal of Experimental Botany. 62:131-135.

Ramírez, L.R.G., Neira-Morales R.R. y Torres Noriega, J.A. 2000. Digestión ruminal de la proteína de siete arbustos nativos del noreste de México. International Journal of Experimental Botany. 67:29-35.

Ramírez, R.G. 1998. Nutrient digestion and nitrogen utilization by goats fed native shrubs Celtis palida, Leucophullum texanum and Porlieria angustifolia. Small Ruminant Research. 28:47-51.

Ramirez, R.G. 1999. Feed resources and feeding techniques of small ruminants under extensive management conditions. Small Ruminant Research, 34(3): 215-230.

Ramírez, R.G. and Ledezma-Torres, R.A. 1997. Forage utilization from native shrubs Acacia rigidula and Acacia farnesiana by goats and sheep, Small Ruminant Res., 25: 43-50.

Ramírez, R.G., González Rodríguez, H., Gómez-Mesa, M. y Pérez-Rodríguez, M.A. 1999. Feed value of foliage from Acacia rigidula, Acacia farnesiana and Acacia berlandieri, J. Appl. Anim. Res., 16: 23-32.

Ramírez, R.G., Neira-Morales, R.R., Ledezma-Torres, R.A., Garibaldi-González, C.A. 2000. Ruminal digestion characteristics and effective degradability of cell wall of browse species from northeastern Mexico. Small Ruminant Research, 36(1): 49-55.

Ramírez, R.G., Neira-Morales, R.R., Torres-Noriega, J. and Mercado-Santos, A.C. 2000c. Seasonal variation of chemical composition and crude protein digestibility in seven shrubs of NE México. FYTON, Journal of Experimental Botany, 68:77-82.

Ramírez, R.G., Neira-Morales, R.R., Torres-Noriega, J.A. 2000b. Digestión ruminal de la proteína de siete arbustos nativos del nordeste de México. International Journal of Experimental Botany, FYTON, 67(1): 29-35.

Ramírez, R.G., Neira-Morales, R.R., Torres-Noriega, J.A., Mercado-Santos, A.C. 2000a. Seasonal variation of chemical composition and crude protein digestibility in seven shrubs of NE Mexico. International Journal of Experimental Botany, FYTON, 67(2): 77-82.

Ramírez, R.G., R.R. Neira-Morales, R.A. Ledezma-Torres, C.A. Garibaldy-González. 2000. Ruminal digestion characteristics and effective degradability of cell wall of browse species from northeastern Mexico. Small Ruminant Research. 36: 49-55.

Ramírez, R.G., R.R. Neira-Morales, y J.A. Torres-Noriega. 2000b. Digestión ruminal de la proteína cruda de siete arbustos nativos del noreste de México. International Journal of Experimental Botany. FYTON, 67:29-35.

Ramírez, R.G., Sauceda, J.G., Narro, J.A. and Aranda, J. 1993. Preference indices for forage species grazed by Spanish goats on a semiarid shrubland in México. J. Appl. Anim. Res 3: 55-66.

5. Ramírez, R.G. 1999. Feed resources and feeding techniques of small ruminants under extensive management conditions. Small Ruminant Research. 34: 215-230.

Ramírez-Lozano, R.G. 2004. Nutrición del Venado Cola Blanca. Publicaciones Universidad Autónoma de Nuevo León, San Nicolás de los Garza, N.L., México.

Ramírez-Lozano, R.G. 2004. Nutrición del Venado Cola Blanca. Universidad Autónoma de Nuevo León Press, San Nicolás de los Garza, N.L., México: pp 133-154.

Ramírez-Lozano, R.G. 2006. Nutritional characteristics of browse species from Northeastern Mexico consumed by small ruminants. BSAS Publication 34. The assessment of intake, digestibility and the roles of secondary compounds. Edited by C.A. Sandoval-Castro, F.D.DeB.D. Hovell, J.F.J. Torres-Acosta and A. Ayala-Burgos. Nottingham University Press. pp. 251-260.

Ramírez-Lozano, R.G. 2009. Nutrición de Rumiantes: Sistemas Extensivos. Editorial Trillas, 2ª edición. México, pp. 221-265.

Ramírez-Lozano, R.G. 2011. Browse Foliage as Protein Supplement for Sheep Fed Low Quality Diets. Edit. LAP Lambert Academic Publishing, Germany, pp. 37-85.

Ramirez-Lozano, R.G., Alvarado M. del S. y González-Rodríguez, H. 2010b. Mineral Content in Browse Plants. LAP Lambert Academic Publishing, Germany, pp. 32-57.

Ramírez-Lozano, R.G., González-Rodríguez, H., Gómez-Mesa, M.V., Cantú-Silva, I. and Uvalle-Sauceda, J.I. 2010a. Spatio-temporal variations of macro and trace mineral contents in six native plants consumed by ruminants at northeastern Mexico. Journal of Tropical and Subtropical Agroecosystems, 12(2): 267-281.

Ramirez-Lozano, R.G., H. Gonzalez-Rodriguez, M.V. Gómez-Meza, I. Cantú-Silva, and J.I. Uvalle-Sauceda (2010). Spatio-temporal variations of macro and trace mineral contents in six native plants consumed by ruminants at northeastern México. J. Trop. Subtrop. Agroecosyst. 12:267-281.

Ramírez-Orduña, R., R. G. Ramírez, M. V. Gómez-Meza, J. A. Armenta-Quintana J. M. Ramírez-Orduña, R. Cepeda-Palacios J. M. Ávila-Sandoval. 2003. Seasonal dynamics of crude protein digestion in browse species from Baja California Sur, México, Interciencia, 28(3): 408-414.

Rangel Rosete, Ana Luisa. Industrializacion del Brasil (Condalia hookeri Johnst) y coma (Bumelia celastrina H.B.K.) como frutos silvestres en Marin, N.L. p 26.

Raven, P.H., Evert, R.F. & Eichhorn S.E. 2005: Biology of Plants (7thEdition). W.H. Freeman & Company. New York.

Rea, Amadeo M. 1991. Gila River Pima dietary reconstruction. Arid Lands Newsletter. 31: 3-10.

Reed, J. D. 1995 Nutritional toxicology of tannins and related polyphenols in forage legumes. Journal of Animal Science, 73:1516-1528

Reyes, M.J.L. y Ortega, S.J.L. 1994. Chamizo (Atriplex canescens) como sustituto parcial de alfalfa (Medicago sativa) en dietas para cabritos en desarrollo. Tesis. Universidad Autónoma Chapingo, Unidad Regional Universitaria de Zonas Áridas, Bermejillo, Durango.

Reynel, C. 1995. Syst. Neotrop. Zanthoxylum 1–657. Unpublished Ph.D. thesis, University of Missouri---St. Louis, St. Louis. Wunderlin, R. P. 1998. Guide Vasc. Pl. Florida i–x + 1–806. University Press of Florida, Gainseville.

Reynolds, C. K., H. F. Tyrrell, and P. J. Reynolds. 1991. Effects of diet forage-to-concentrate ratio and intake on energy metabolism in growing beef heifers: whole body energy and nitrogen balance and visceral heat production. J. Nutr. 121:994-1003.

Rich, P.R. 2003. The molecular machinery of Keilin's respiratory chain. Biochemical Society Transactions 31: 1095-1105.

Richardson, A. 1995. Plants of the Rio Grande Delta. University of Texas Press. pp. 103-104.

Richardson, C. L. 1990. Factors affecting deer diets and nutrition. College Station, TX: Texas A&M University.

Rittner, U. and Reed, J.D. 1992. Phenolics and in vitro degradability of protein and fibre in West African browse. Journal of the Science of Food and Agriculture, 58: 21–28.

Robbins, C. T. 1993. Wildlife feeding and nutrition. 2nd ed. New York, NY: Academic Press. 352 p.

Robbins, C.T., T.A. Hanley, A. E. Hagerman, O. Hjeljord, D.L. Baker, C.C. SCHWARTZ, and W W. Mautz. 1987. Role of tannins in defending plants against ruminants: reduction in protein availability. Ecology 68:98–107.

Robertson, J. B. y P. J. Van Soest. 1991. The detergent system of analysis and its application to human foods. In: The analysis of dietary fiber in food. Marcel Dekker, Inc. p 123-156.

Rodriguez-Santillan, P., Bernal-Barragan, H., Cerrillo-Soto, M., Gonzalez-Rodriguez, H., Juarez-Reyes, A. S., Guerrero-Cervantes, M., & Ramirez-Lozano, R. G. (2014). Leaf litter as a food resource for range livestock. JAPS, Journal of Animal and Plant Sciences, 24(6): 1629-1635.

Rodríguez, D.R. 2010. Consumo de hojas jóvenes de roble (Quercus pyrenaica) por el ganado vacuno: aspectos nutricionales e intoxicación. Memoria de Tesis Doctoral. Instituto de Ganadería de Montaña. Consejo Superior de Investigaciones Científicas (CSIC). Universidad de León.

Rojas Hernandez S., Quiroz Cardoso F., Camacho Diaz L.M., Cipriano Salazar M., Avila Morales B., Cruz Lagunas B., Jimenez Guillen R., Villa Mancera A., Abdelfattah Z.M. Salem. and Olivares Pérez J. 2015. Productive Response and Apparent Digestibility of Sheep Fed on Nutritional Blocks with Fruits of Acacia farnesiana and Acacia cochliacantha. Life Science Journal, 2015: 12(2s).

Romero L C E, Palma G J M y López J 2000 Influencia del pastoreo en la concentración de fenoles y taninos condensados en Gliricidia sepium en el trópico seco. Livestock Research for Rural Development 4(12):1-9 http://www.cipav.org.co/lrrd/lrrd12/4/rome124.htm.

Sadleir, R.M.F.S. 1982. Energy consumption and subsequent partitioning in lactating black-tailed deer. Canadian Journal of Zoology 60:382–386.

Safari J, Mushi DE, Kifaro GC, Mtenga LA, Eik LO. 2011. Seasonal variation in chemical composition of native forages, grazing behavior and some blood metabolites of Small East African goats in a semi-arid area of Tanzania. Animal Feed Science and Technology 164:62-70.

Sakuragi, M., H. Igota, H. Uno, K. Kaji, M. Kaneko, R. Akamatsu, and K. Maekawa. 2003. Seasonal habitat selection of an expanding sika deer Cervus nippon population in eastern Hokkaido, Japan. Wildlife Biology, 9:141–153.

Salem, B.H., Abidi, S., Makkar, H.P.S. and Nefzaoui A. 2005. Wood ash treatment, a cost-effective way to deactivate tannins in *Acacia cyanophylla* Lindl. foliage and to improve digestion by Barbarine sheep. Animal Feed Science and Technology 123: 93-108.

Salem, A.Z.M. et al. 2007. In vitro fermentation and microbial protein synthesis of some browse tree leaves with or without addition of polyethylene glycol. Anim. Feed Sci. and Technol. 138(3): 318-330.

Salinas, H., Hoyos, G. y Sáenz, P. 1989. El sistema de producción caprino en la comarca lagunera. En: Taller de trabajo: sanidad y reproducción de caprinos. Edit. H. Salinas, S. Flores y F. Ruiz. Centro Internacional de Investigaciones para el Desarrollo. INIFAP. SARH. Matamoros, Coahuila. México.

Sallam, S.M.A., Bueno, I.C.S., Godoy, P.B., Nozella, E.F., Vitti, D.M.S.S., Abdalla, A.L., 2010. Ruminal fermentation of tannins bioactivity of some browses using a semiautomated gas production technique. Trop. Subtrop. Agroecosyst. 12:1-10.

Sánchez Sánchez Oscar.1980. Flora del valle de México. Edit. Herrero. 4ta edición. P: 193.

Sandoval-Castro, C.A., Torres-Acosta, J.F.J., Hosteb, H., Salem, A.Z.M., Chan-Pérez, J.I., 2012. Using plant bioactive materials to control gastrointestinal tract helminths in livestock. Anim. Feed Sci. Tech. 176:192-201.

Sangines, I., Grande, D. y Perez-Gil, F. 1992. Comparación del valor nutritivo de cuatro especies de atriplex y evaluación de un procedimiento de desalado sobre el contenido proteínico de A. nummularia., Biotam. 4:56-62.

Shelton, H.M. 2004. Importance of tree resources for dry season feeding and the impact on productivity of livestock farms. In: 't Mannetje L, Ramirez L, Ibrahim M, Sandoval C, Ojeda N, Ku J (eds) The importance of silvopastoral systems in rural livelihoods to provide ecosystem services. Proceedings of 2nd International Symposium on Silvopastoral Systems, Universidad Autónoma de Yucatán, Mérida,Yucatán, México.

Schoenherr, A.A. 1995. A Natural History of California. Berkeley, CA: University of California Press, p. 14.

Schulman, R.S. 1992. Statistics in plain English. New York, NY: Chapman and Hall.

Scifres, C. J. 1980. Brush management, principles and practices for Texas and the Southwest. College Station, TX: Texas A&M University Press. 360 p.

Secretaría de Salud. 1993. Edición conmemorativa. La investigación científica de la herbolaria medicinal mexicana. México, pp. 209-218.

Shenkute, B., Hassen, A., Assafa, T., Amen, N. and Ebro, A. 2012. Identification and nutritive value of potential fodder trees and shrubs in the Mid Rift Valley of Ethiopia, The Journal of Animal and Plant Sciences, 22: 1126-1132.

Singh, B., Makkar, H.P.S. and Negi, S.S., 1989, Rate and extent of digestion and potentially digestible dry matter and cell wall of various tree leaves. J. Dairy Sci., 72:3233-3239.

Singh, S., Kushwaha, B.P., Naga, S.K., Mishra, A.K., Singh, A., Anelec, U.Y., 2012. In vitro ruminal fermentation, protein and carbohydrate fractionation, methane production and prediction of twelve commonly used Indian green forages. Anim. Feed Sci. Tech. 178:2-11.

Sirmah, P., Muisu, F., Mburu, F., Dumarçay, S., & Gérardin, P. 2008. Evaluation of Prosopis juliflora properties as an alternative to wood shortage in Kenya. Bois et Forêts des Tropiques, 298(4), 25-35.

Smith, G. S. 1992. Toxification and detoxification of plant compounds by ruminants: an overview. Journal of Range Management 45:25–30.

Smith, J.K., J.P. Neel, and E.D. Felton (2009). Utilization of leaf litter as a potential feed source. (abstract). In: Proceedings of the American Society of Animal Science, Midwestern Section Annual Meeting, March 16-18, Des Moines, IA. USA, pp. 136-138.

Smith, S.H., J.B. Holter, H.H. Hayes, and H. Silver. 1975. Protein requirements of white-tailed deer fawns. Journal of Wildlife Management 39:582–589.

Soltero, S. y Fierro, L.C. 1980. Contenido y fluctuación nutricional de chamizo (Atriplex canescens) durante el periodo de sequía en un matorral micrófilo. Boletín Pastizales, R.E.L.C.-I.N.I.P. S.A.R.H. Vol II (5).

Soltero, S. G. y Fierro, L.C. 1981. Importancia del chamizo (Atriplex canescens) en la dieta de bovinos en pastoreo en un matorral desértico de Atriplex-Prosopis durante la época de sequía. Boletín Pastizales, RELC, INIP, SARH, Vol. XII No.1.

Sosa R E E, Pérez R D, Ortega R L y Zapata B G 2004 Evaluación del potencial forrajero de árboles y arbustos tropicales para la alimentación de ovinos. Técnica Pecuaria en México. 42(2):129-144.

Spalinger, D.E., S.M. Cooper, D.J. Martin, and L. A. Shipley. 1997. Is social learning an important influence on foraging behavior in white-tailed deer? Journal of Wildlife Management 61:611–621.

Spears, J.W. 1994. Minerals in forages. In: Fahey Jr. GC. (Ed.). National Conference on Forage Quality, Evaluation and Utilization. University of Nebraska, Lincoln, N.E., USA, pp. 281-317.

Spears, J.W. 1998. Reevaluation of metabolic essentiality of minerals. Proceedings: New Technologies for the Production of "Next Generation" Feeds and additives. The 8th World Conference on animal Production, Seoul National University, Seoul Korea, pp. 68-77.

SPP–INEGI (1986). Sintesis Geografica del estado de Nuevo Leon. Secretaría de Programación y Presupuesto, Instituto Nacional de Geografía e Informática, México, pp. 17-19.

Springfield, H.W. 1970. Germination and stablishment of fouring saltbush. In: The south west USDA Serv. Res. Pap. R.M. 55

Standley, P.C. 1992. Contributions from the National Herbarium. Trees and shrubs of Mexico. Washington, DC: Smithsonian Institution. 23: 1721-1725.

Steinshamn, H. 2010. Effect of forage legumes on feed intake, milk production and milk quality: a review. Animal Science Papers and Reports, 28: 195-206.

Stewart, Donovan. 1970. Manual of the Vascular Plants of Texas. Texas Research Foundation Editorial. Texas, p.p. 942-943

Stienen, H., M.P. Smits, N. Reid, J. Landa, and J.H.A. Boerboom (1989). Ecophysiology of 8 woody multipurpose species from semiarid northeastern Mexico. Ann. Sci. Forest. 46:454-458.

Stuth, J.W., Lyons, R.K. 1999. Grazing steer fecal output dynamics on south Texas shrub-land. Journal of Range Management, 52(3): 275-282.

Sultan, S., Negi, A.S., Agarwal, D.K., Katiyar, P.K. y Singh, U.P. 2000. Chemical composition, in vitro dry matter digestibility, total phenolics and proanthocyanidins in hedge lucerne (Desmanthus virgatus) exotic germplasm. Indian Journal of Animal Sciences, 70: 1246-1249.

Tello Sanchez, Daniel. 1989. Estudio fenologico de la coma (Bumelia celastrina H.B.K.) en 4 municipios del estado de Nuevo Leon, pp. 4-11.

Thies, Monte; Caire, William. 1990. Association of Neotoma micropus nests with various plant species in southwestern Oklahoma. Southwestern Naturalist. 35(1): 80-102.

Tine, M. A., K. R. McLeod, R. A. Erdman, and R. L. Baldwin VI. 2001. Effects of brown midrib corn silage on the energy balance of dairy cattle. J. Dairy Sci. 84:885-895.

Turner, I. M. (1994). Sclerophylly: Primarily protective? Functional Ecology 8, 669-675.

Umachigi, S.P., K.N. Jayaveera, C.K. Ashok Kumar, G.S. Kumar, B.M. Vrushabendr swamy y D.V. Kishore Kumar. 2008. Studies on Wound Healing Properties of Quercus infectoria. Trop. J. Pharm Res. 7 (1).

Underwood, E.J., N.F. Shuttle. 1999. The Mineral Nutrition of Livestock, 3rd edn. CAB International, Wallingford, UK, pp. 100-200.

Valencia, C.M. 1991. Utilización de Atriplex canescens (Pursh) Nutt. En: Memoria del taller sobre captación y aprovechamiento del agua con fines agropecuarios en zonas de escasa precipitación. Eds. H. Salinas, S. Flores, M. Martínez. INIFAP. SARH. p 255-270.

Valencia, C.M. y Nava, T. 1981. Época y frecuencia de utilización de Atriplex canescens (Pursh) Nutt. Univ. Autónoma Agraria Antonio Narro. Monog. técnico-científica. 7 (1) 1-67.

Valencia, A. S. 2004. Diversidad del género Quercus (Fagaceae) en México. Bol. Soc. Bot. México 75:33-53.

Van Soest, P. J. 1994. Nutritional Ecology of the ruminant. Second edition. Editional Comstock Publishing associates, Cornell University Press. Pp. 312-317.

Van Soest, P.J. 1982. Nutritional ecology of the ruminant. Corvallis, OR: O & B Books. 374 p.

Van Soest, P.J. 1993. Cell wall matrix interactions and degradation-session synopsis. In Forage cell wall structure and digestibility, Editors: H.G. Jung, D.R. Buxton, R.D. Hatfield and J. Ralph. p 377. ASA. CSSA-SSSA, Madison, WI, USA.

Van Soest, P.J. 1994. The Nutritional Ecology of the Ruminant 2nd ed. Commstock Publisher Associated. Cornell University Press. pp. 312-317.

Van Soest, P.J., J.B. Robertson, and B.A. Lewis (1991). Methods for dietary, neutral detergent fiber, and nonstarch polysaccharides in relation to animal nutrition. Symposium: carbohydrate methodology, metabolism, and nutritional implications in dairy cattle. J. Dairy Sci. 74:3583-3597.

Varner, L.W., and L.H. Blankenship. 1987. South Texas shrubs: nutritive value and utilization by herbivores. Proceedings of the Symposium on Plant Herbivore Interactions. USDA Forest Service. USDA Forest Service General Technology Report INT-222. p 108–112.

Varner, L.W., L.H. Blankenship, and G.W. Lynch. 1977. Seasonal changes in nutritive value of deer food plants in south Texas. Proceedings of the Annual Conference of the Southeastern Association of Fish and Wildlife Agencies 31:99–106.

Vázquez, M. 1981. Determinación de la dieta de caprinos en un matorral desértico microfilo del municipio de Ocampo, Coahuila. México. Tesis. Universidad Autónoma del Noreste, Saltillo, Coah.

Vazquez, R., C. Gallegos, N.E. Treviño, Y. Díaz, 1998, Congreso sobre conocimiento y aprovechamiento del Nopal, VII Congreso Nacional y V Congreso Internacional, Monterrey, N.L. México, pp. 82-86.

Ventura, M.R., Castanon, J.I.R., Rey, L., Flores, M.P., 2002. Chemical composition and digestibility of Tagasaste (Chamaecytisus proliferus) subspecies for goats. Small Ruminant Research, 46(2): 207-210.

Villa, S. A. 1981. Los Desiertos de México. In: General Technical Report WO-28 Arid Land Resource Inventories: Developing Cost-Efficient Methods. An International Workshop November 30 December 6, 1980. La Paz, México, pp. 18-20.

Villee, C.A., Solomon, E.P., Martin, C.E., Martin, D.W., Berg L.R. and Davis, R.W. 1989. Biology (2nd Edition). Saunders College Publishing. Fort Worth.

Vines, R. A. 1982. Trees of North Texas. University of Texas Press, Austin. 1stedition. p.p. 252-257.

Vines, R. A. 1984. Trees, shrubs, and woody vines of the Southwest. Austin, TX: University of Texas Press. 405 p.

Vines, R.A. 1960. Trees, shrubs and woody vines of the southwest. U. of Texas Press. Austin, Texas. USA.

Vora, Robin S. 1990. Plant phenology in the lower Rio Grande Valley, Texas. Texas Journal of Science. 42(2): 137-142.

Waghorn, G. 2008. Beneficial and detrimental effects of dietary condensed tannins for sustainable sheep and goat production. Progress and challenges Animal Feed Science and Technology, 147: 116-139.

Waghorn, G.C. and W.C. McNabb. 2003. Consequences of plant phenolic compounds for productivity and health of ruminants. Proc. Nutr. Soc. Cambridge University Press. 62:383-392.

Walter, K. J., & Armstrong, K. V. 2014. Benefits, threats and potential of Prosopis in South India. Forests, Trees and Livelihoods, 23(4), 232-247.

Ward, R. L. and C. L. Marcum. 2005. Lichen litterfall consumption by wintering deer and elk in western Montana. Journal of Wildlife Management 69:1081-1089.

Wasowski, S. and A. Wasowski. 2002. Native texas plants: landscaping region by region Publisher: Lanham, Md. and Lone Star Books, USA.

Whitehead, D.C. 2000. Nutrient Elements in Grassland: Soil-Plant-Animal Relationships. CABI, Publishing. UK, pp. 150-162.

Woodcock, B.G., and C.C. Wood. 1971. Effect of protein-free diet on UDP glucuronyltransferase and sulphotransferase activities in rat liver. Biochemical Pharmacology 20:2703–2713.

Yayneshet, T., Eik, L.O., Moe, S.R. 2009. Seasonal variations in the chemical composition and dry matter degradability of exclosure forages in the semi-arid region of Northern Ethiopia. Animal Feed Science and Technology, 148(2):12-33.

Youssef, F.G. 1988. Some factors affecting the mineral profiles of tropical grasses. Outlook on Agriculture, 17; 104-111.

Zuloaga, F. O., O. Morrone, M. J. Belgrano, C. Marticorena & E. Marchesi. (eds.) 2008. Catálogo de las Plantas Vasculares del Cono Sur (Argentina, Sur de Brasil, Chile, Paraguay y Uruguay). Monogr. Syst. Bot. Missouri Bot. Gard. 107(1): i–xcvi, 1–983; 107(2): i–xx, 985–2286; 107(3): i–xxi, 2287–3348.

VITA

Roque G. Ramírez-Lozano

Address:

Universidad Autónoma de Nuevo León, Facultad de Ciencias Biológicas, Alimentos. Ave. Pedro de Alba y Manuel Barragán S/N, Ciudad Universitaria, San Nicolás de los Garza, Nuevo León, 66455, México.

Education:

B.Sc. in Agriculture, Universidad Autonoma de Nuevo Leon in 1972; M.Sc. in Animal Science, New Mexico State University in 1983; Ph.D. in Animal Nutrition, New Mexico State University in 1985.

Present Position:

Professor, Department of Food Sciences.

Job Duties:

Professor-researcher. Undergraduate and graduate teaching in Animal Nutrition, Statistics and Research Techniques.

Research Awards:

Sistema Nacional de Investigadores, Nivel Tres

Graduate Student Supervision:

101 undergraduate students, 14 M.Sc. students and 24 Ph.D. students.

Publications:

148 peer-reviewed journal articles, 35 invited papers, 11 books, 7 book chapters, 62 published proceedings articles and 60 abstracts.

Published works have been cited 1515 times as of August 2015, Index h 22 and index i10 55.

Editorial and Professional Service:

Journal of Animal Science, International Goat Association, several Mexican Animal Science Associations